Lehrstuhl für Hochfrequenztechnik
Technische Universität München

Resolution and Accuracy in Multiband Radar Sensors

Simón Tejero Alfageme

Vollständiger Abdruck der von der Fakultät für Elektrotechnik und
Informationstechnik der Technischen Universität München zur Erlangung des
akademischen Grades eines

Doktor-Ingenieurs

genehmigten Dissertation.

Vorsitzender: Univ.-Prof. P. Lugli, Ph. D.
Prüfer der Dissertation: 1. Univ.-Prof. Dr.-Ing., Dr.-Ing. habil. J. Detlefsen
                             2. Univ.-Prof. Dr.-Ing. W. Utschick

Die Dissertation wurde am 23. November 2006 bei der Technischen
Universität München eingereicht und durch die Fakultät für Elektrotechnik
und Informationstechnik am 24. Mai 2007 angenommen.

Bibliografische Information der Deutschen Nationalbibliothek

Die Deutsche Nationalbibliothek verzeichnet diese Publikation in der
Deutschen Nationalbibliografie; detaillierte bibliografische Daten sind
im Internet über http://dnb.d-nb.de abrufbar.

ISBN 978-3-8325-1661-1

Logos Verlag Berlin
Comeniushof, Gubener Str. 47,
10243 Berlin
Tel.: +49 030 42 85 10 90
Fax: +49 030 42 85 10 92
INTERNET: http://www.logos-verlag.de

a Ana

Acknowledgment

This thesis was realized during my research activities in the *Lehrstuhl für Hochfrequenztechnik* of the *Technische Univsersität München*. I would like to thank Univ.-Prof. Dr. techn. Dr. h.c. Peter Russer for accepting me in the institute.

My advisor, Univ.-Prof. Dr.-Ing. Jürgen Detlefsen, also deserves credit. He not only provided me with an excellent topic, but he also patiently allowed me both the freedom and the time needed to bring this work into full fruition. I consider it an honor to have worked for him.

Univ.-Prof. Dr.-Ing. Wolfgang Utschick and Univ.-Prof. Dr. Paolo Lugli graciously agreed to serve on the committee and review this work. It is professors such as these that make the *Technische Universität München* the great university that it is.

I have also benefited over the years from the excellent atmosphere in *Fachgebiet Hochfrequente Felder und Schaltungen* under the direction of Univ.-Prof Dr.-Ing Jürgen Detlefsen. To him and to my colleagues with whom I have had the pleasure to work with, I am humbly indebted. To Uwe Siart, my project partner, with whom I spent many hours discussing about dualband radar. He also provided together with Alexander Dallinger and Simon Schelkshorn the computer environment that allowed me a smooth work. Especially I would like to thank Volker Winkler, my office colleague and friend, for reviewing this work, as well as for his advices and friendship.

I want to express my gratitude to my great friend Daniel Beltran, who helped me with some grammatical aspects of this thesis.

This work would had not been possible without the education I received during my School time in *Colegio San Alberto Magno* and *I. B. Félix de Azara* in Zaragoza (Spain), where I was not only introduced to scientific work, but also educated on solid principles. I would also like to thank the *Universidad de Zaragoza*, where I was introduced to Electrical Engineering.

I really appreciate the support I received from my wife Ana, to whom I owe so much. Of course not least, my gratitude goes to my family and friends, without whom all this would not be possible.

Munich, July 2007

Abstract

This thesis deals with the jointly processing of radar signals stemming from independent sensors observing the same scattering scenario at different frequency bands. The suitability of a simple point-scatterer model to describe complex scattering mechanisms is verified over wide bandwidths. The minimum error in the estimation of multiband radar signal parameters is obtained using the Cramer-Rao theorem. The resolution problem (separation of targets) is treated as a test of the hypotheses of one or two closely-spaced scatterers present in the radar signal. Accuracy and resolution limits for multiband data sets are obtained for different distances between the frequency bands. To process multiband radar signals, high-resolution spectral estimators are adapted to cope with the multiband case. While the accuracy improves with increasing band distance, no general improvement of the resolution performance is observed. Additionally, an approach to estimate the distance between the frequency bands is developed. Monte Carlo analyses confirm the fundamental behavior of the resolution and accuracy dependence on the band gap.

Zusammenfassung

Die gemeinsame Verarbeitung von Radarsignalen unabhängiger Sensoren, die das gleiche Streuszenario in unterschiedlichen Frequenzbereichen beobachten, wird untersucht. Ein einfaches Punktstreuermodell zur Beschreibung komplexer Streumechanismen über große Bandbreiten hinweg wird überprüft und der minimale Schätzfehler für Radarsignalparameter bei Messung in zwei Frequenzbändern ermittelt. Das Auflösungsproblem (Zieltrennung) wird als Test von Hypothesen über das Vorhandensein von ein oder zwei benachbarten Streuzentren behandelt und Ortungsgenauigkeit und Auflösungsgrenzen für Daten aus zwei Frequenzbändern bei unterschiedlichem Frequenzbandabstand bestimmt. Die Mehrband-Signalauswertung erfolgt mit hochauflösenden spektralen Schätzverfahren, die entsprechend erweitert werden. Während die Ortungsgenauigkeit mit zunehmendem Bandabstand zunimmt, ergibt sich keine generelle Verbesserung der Auflösung. Zusätzlich wird ein Ansatz zur Schätzung des Frequenzbandabstandes entwickelt. Monte Carlo Analysen bestätigen grundsätzlich die gefundenen Abhängigkeiten.

Contents

1. Introduction

In the last years, the use of radar systems to sense and measure the environment has experienced a great growth, especially beyond its traditional area, the military applications. The accuracy and reliability of the information obtained from the environment will obviously depend on the quality and diversity of the available data. Data quality can be assessed using the energy relation between the information data and the different nuisance signals present in it. Diversity applies to the variety and amount of information collected by the radar sensor, e. g. frequency, aspect angle or polarization. In inverse scattering problems, different sensing aspect angles are unavoidable for object shape estimation. In range estimation problems, where the interest of this thesis is focused on, it is a well known fact that range accuracy and resolution, i. e. exactitude in the estimation and the capability to detect two closely spaced objects as two independent ones, depend mostly on the total signal bandwidth that can be used for range profile estimation. The frequency bands available for concrete applications, however, are usually restricted in frequency bandwidth and as a result, a limited, maybe insufficient, range resolution and accuracy can be obtained.

To increase the diversity, many radar sensors covering different aspect angles, frequency bands or wave polarizations can be used. The general case of a sensor platform is shown in figure 1.1. A set of radar sensors observes a common scenario under different aspect angles and operating at different frequency bands $\mathcal{B}_l$. Nowadays, such platforms become available. Reduction of production costs, better integration and increased demand on sensing functions will enhance this trend in the future. The most popular example for the expansion of different radar applications in a civil platform is the automobile. More and more comfort and security applications as autonomous cruise control, parking aid, blind-spot surveillance or pre-crash warning, require a reliable and accurate sensing of the car environment. Determined by international regulators, frequency bands at $24\,\mathrm{GHz}$, $60\,\mathrm{GHz}$, $77\,\mathrm{GHz}$ and $142\,\mathrm{GHz}$ can be used for this services. Not only in automotive applications, also in services like remote earth sensing, weather monitoring, air-traffic control or ballistic defense different radar sensors operating at diverse frequency bands are used. The joint use of all the available information from sensors working at different frequency bands can be used to increase the quality and reliability of the environment measurement.

Nowadays, sensor diversity is usually employed at the data level, i. e. *data fusion* techniques are applied (see, e. g. [26]). Data which have been already

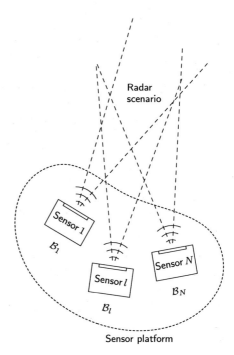

Figure 1.1: General platform of radar sensors working at different frequency bands $\mathcal{B}_l$ observing a common radar scenario

Sensor platform

processed by the individual sensors in the platform, are used jointly to better detect or classify an object or to increase reliability. Range profile details which are not resolved by the sensors independently can not be resolved after the data fusion process either. A further exploitation of the sensor diversity can be, therefore, achieved by jointly processing the data at the RF level. The full diversity, e.g. full signal bandwidth, can be exploited to increase the sensing performance.

An important issue in multisensor platforms is the *coherence* between the diverse radar sensors. A coherent system implies an exact knowledge of the relative positions between the diverse frequency bands in order to be able to use the phase information. This frequency distance control requires a higher hardware cost, so that algorithms, which are able to jointly process data streams stemming from *non-coherent* sensors, are of concern.

Range profile estimation from bandwidth-limited frequency domain reflection measurements is strongly related to spectral estimation from time-limited data sets. In both cases, the available data and the desired representation are related to each other by the Fourier transform. The Fourier transform is found in many other signal processing problems, e.g. direction of arrival estimation of single-frequency plane waves with antenna arrays. Algorithms, statements and results found for one case are applicable to the others.

1.1. State of the art

The increase in signal bandwidth to enhance resolution capabilities has been directly and indirectly used in a variety of classic radar systems [16, 49, 52, 55, 59, 63, 83, 84].

In pulsed radars, range is determined by measuring the round-trip travel time of a transmitted pulse to a scatterer and back to the antenna. The minimum distance of two targets to be detected independently, i. e. the resolution distance, is given by the pulse width. The resolution, therefore, is enhanced by decreasing the pulse width, which also means an increase of the signal bandwidth. The detection capabilities depend directly on the energy of a single pulse, which in peak-power-limited systems is determined by the pulse duration. It is impossible, therefore, to increase both parameters at the same time. A trade-off between detectability and resolvability must be established. Pulse compression techniques are intended to break this trade-off. The pulse autocorrelation width, i. e. the resolution distance for optimum receivers, is reduced without decreasing the pulse duration by means of pulse modulation. This modulation implies, as expected, an increase in signal bandwidth.

In *continuous wave* (CW) radars, one single carrier at a constant frequency is continuously transmitted. These systems, based on the Doppler principle, are designed to measure speed. The incapability to measure range is concordant with the infinitesimally small bandwidth occupied by the transmitted signal. In *frequency modulated continuous wave* (FMCW) radars, the carrier is frequency modulated, e. g. by a triangular signal. The difference between the transmitted and received frequencies is proportional to the target range. The resolution distance is given by the capability to detect independently two closely spaced frequencies and depends in this case on the maximum instant frequency deviation or modulation range, i. e. again on the bandwidth of the transmitted signal.

To increase the frequency diversity, multicarrier signals can be used. A *multiple-frequency continuous wave* radar, also called *frequency shift keying continuous wave* (FSK-CW), uses two different unmodulated carriers to measure distance and speed of one object [16, 83, 84]. An unmodulated carrier reflected on a target suffers a frequency shift proportional to the target speed, and a phase shift proportional to its range. This range measure, however, has a very short *unambiguous range*, i. e. the maximum distance which can be assigned to a target unambiguously. By evaluating the phase shift at two different carrier frequencies, a measure of the target distance with an unambiguous range proportional to the inverse of the frequency distance can be obtained. With increasing frequency distance, a better range accuracy is achieved. A trade-off between the accuracy and unambiguous range must be established. This system can only resolve targets based on its Doppler shift, i. e. no resolution based on range measurements is actually achieved. This is concordant with the zero bandwidth of the transmitted signal, as only two spectral lines are present. The trade-offs and statements found in this system will also be found in more general multiband systems. The resolution capability will be given by the total bandwidth, and increasing the distance between the bands will imply a better accuracy but acuter ambiguity problems.

More general multicarrier radar signals have been proposed recently, the so-called *multicarrier phase-coded* (MCPC) signals [50, 51, 52]. Every MCPC pulse consists of N sequences (phase codes) transmitted simultaneously on N carriers. Each sequence contains M phase-modulated bits. The frequency difference between two adjacent carriers is set equal to the inverse of the bit duration yielding to *orthogonal frequency division multiplexing* (OFDM). This kind of signals allows a very good control of the signal bandwidth and the ambiguity function, which is a measure of the resolution capability of a radar signal. As disadvantage, a strong amplitude modulation appears in the transmitted signal, which limits the kind of energy efficient power amplifiers that can be used.

A multiband frequency ultra-wideband CW radar system has been recently proposed for mine detection [23, 24, 33]. The so-called PANDORA radar is a multiband radar which subdivides the bandwidth in eight channels that work at the same time. The simultaneous sensing at different channels reduces the scanning time. Usually, the frequency bands are adjacent, so that results for general modulated CW radars can be applied. An approach considering guard bands, i. e. not used regions between the frequency channels, is considered in [32], where an analysis of the performance based on the ambiguity function is presented. Due to the increased synthetic bandwidth obtained with the guard bands, a narrower response is obtained in the delay axis and, following classic radar signal theory, a better resolution is achieved. The response presents also several peaks with high amplitude, which lead to ambiguity problems.

In the multiband approaches described previously, the different frequency bands were adjacent to each other or, as in the last case, defined guard bands were present. The interest of this thesis focuses on radar systems with an arbitrary gap between the frequency bands and on the possibilities to enhance the radar performance by exploiting this band gap. An interesting approach to process data obtained by independent sensors working at different frequency bands has been presented by Cuomo *et al.* [14]. It is based on the decomposition of the frequency-domain radar response in a sum of complex exponentials, a so-called pole model. A modified Root-MUSIC algorithm—a high-resolution spectral estimator—is applied to each subband to obtain a set of pole estimates, which afterwards are aligned in the time domain to correct possible delay errors. This is called by Cuomo a coherent data set, in opposite to the criteria used in this thesis, where global time misalignments are called calibration errors. After time alignment, Cuomo proposes the application of a Root-MUSIC algorithm to obtain a global pole set, which is optimized to reduce the model error over the two subbands. The optimized pole set is used to interpolate the missing frequency response between the two subbands. The gain in resolution is justified by the decrease of the peak widths in the estimated range profile, which is obtained by Fourier transforming the whole—measured and interpolated—data set. In fact, this is not a valid resolution assessment, as all the information stems from the pole model and no information gain can be achieved in the interpolation. Targets which are not included in the signal model will also not appear in the range profile estimation.

An exhaustive analysis of algorithms and possibilities in dualband radar systems has been recently published by Siart [75]. Based also on a pole signal model for the radar data, techniques to process coherently data collected with non-coherent sensors at different frequency subbands are introduced. The data set is assumed to be calibrated and approaches to correct the lack of coherence are presented. Dualband-adapted spectral estimators give a common pole set, which is afterwards optimized to minimize the global model error over the two subbands. Siart places special emphasis on the theoretical background of multiband radar systems. He uses the *Cramer-Rao lower bound* (CRLB) estimation theory to obtain a resolution performance bound. In dualband systems, a resolution improvement with increasing band gap is predicted. The simulations, however, did not show this resolution enhancement, but an improvement of the range accuracy. A similar statement for the resolution behavior, based also on the CRLB theory, has been published more recently by Smith [85].

Different approaches have been developed to assess the accuracy and resolution of radar signals. The ambiguity function is a widely-used classical tool of radar signal analysis [52, 63, 83, 84]. Assuming optimum reception with a matched filter, the squared amplitude of the filter response as a function of the delay and Doppler shift is called the ambiguity function. The width of this function along the delay-axis gives a measure of the resolution capability of the signal. This bound, however, does not take the *signal-to-noise ratio* (SNR) into account, on which the resolution capability depends. To quantify the effect of the SNR, other methods, such as the CRLB, have been proposed.

The CRLB gives the minimum variance for signal parameter estimates which can be obtained with unbiased estimators. The CRLB quantifies a lower bound on estimation error, but it does not indicate directly what is the best resolution achievable by an unbiased estimator. Nevertheless, the CRLB has been widely used to define an absolute limit on resolution [43, 75, 85, 89, 97], which is defined as the source separation that equals its own CRLB or, alternatively, the CRLB of the source estimate.

Zatman and Smith [98] show the incapability of the CRLB to assess the resolution in multiband systems for large band gaps. They analyze the transition for band gaps, where the poles are resolved and the CRLB assumptions are valid, and larger band gaps, where the poles become unresolved and the more general Weiss-Weinstein bound is applied.

In this thesis, the CRLB estimation theory is only used for accuracy analysis. Resolution is considered a *detection* problem, i. e. two targets present in a signal are resolved when both are detected, independent of how accurate they are ranged. Of course, accuracy and resolution are strongly related to each other, but this point of view permits the use of an exact and suitable definition for the resolution event and avoids the use of previous knowledge about the number of sources included in the calculation of the CRLB. Detection or *hypothesis testing* theory is applied to radar signals to obtain optimum discriminators between the single- or dual-target hypotheses. The performance of the optimum discriminators act, analogously as the CRLB for estimation problems, as performance bound. This method results in a novel, algorithm-independent and powerful tool to assess resolution bounds, where an exact and appropriate definition for resolution can be used.

A similar approach using detection theory has been recently applied to resolvability problems in imaging systems [73] and to resolvability of sinusoids in noise [74]. Shahram and Milanfar address in [73] the problem of incoherently detecting two closely spaced light sources observed through a slit. A point source is captured as a spatially extended pattern known as *point-spread function* (PSF), which has, e. g., the form of a sinc-function. Also, they treat in [74] the detection of two closely spaced sinusoids in noise. In both cases, hypothesis testing theory is applied to obtain resolution bounds. Furthermore, the observed PSFs sum—or sinusoids sum—is expanded in Taylor series around the zero source distance to obtain explicit relationships between the minimum detectable source separation and the signal-to-noise ratio. Application or adaptation to interferometric or multiband problems is not treated.

1.2. Aim and contents of this thesis

The aim of this thesis is to analyze the possibilities of increasing accuracy and resolution performance in multiband radar systems by exploiting the frequency band gap. Special attention is given to the non-coherent case, where the band gap between the subbands is only known approximately. Theoretical performance bounds, performance behavior patterns, and suitable signal processing algorithms to exploit the possible performance enhancement are also of concern.

For the analysis, the general multisensor platform depicted in figure 1.1 is simplified to the dualband system in figure 1.2: a platform with two radar sensors observing the same scenario under the same aspect angle and working at different frequency bands $\mathcal{B}_l$. The range profile is assumed to be measured in the frequency domain at discrete frequency points—a *stepped frequency continuous wave* (SFCW) system—so that N_l equally spaced data points are available at each subband. The frequency distance between the two subbands

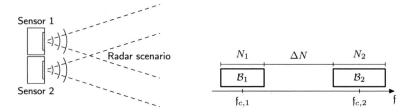

Figure 1.2: Two radar sensors observing the same radar scenario under the same aspect angle and working at different frequency bands $\mathcal{B}_l$

is an arbitrary real number ΔN of data samples. The radar scenario is assumed to be calibrated and static, so that Doppler effects are not taken into account.

To jointly process data collected at different frequency bands, close-to-reality models for the backscattered field of radar scenarios valid for the whole bandwidth should be identified. Chapter 2 presents a signal model for the radar response of an object or a group of them. The scattering response in the frequency domain is modeled by a sum of complex oscillations. Each oscillation can be interpreted as a point scatterer with constant delay and frequency-dependent amplitude. This model is validated with simulated and measured data of canonical and arbitrary-shaped objects.

As it becomes clear from the exponential signal model, target detection from radar signals in the frequency domain is equivalent to oscillation detection and frequency estimation from time-domain signals, so that spectral analysis algorithms can be used for the radar case. Chapter 3 presents a short overview of selected spectral estimators, which are the basis of the multiband algorithms presented later.

Chapter 4 deals with basic theory for accuracy and resolution assessment in radar systems. An overview of classical approaches is given followed by the application to radar signals of the *Cramer-Rao lower bound* (CRLB) and the *hypothesis testing* theory. The CRLB gives a performance bound for range accuracy and hypothesis testing theory is applied to the radar resolution problem. It results in a novel, algorithm-independent and powerful tool to assess resolution bounds.

Theoretical range accuracy bounds in dualband systems are presented in chapter 5. In the non-coherent dualband case, the joint estimation of the radar parameters and the band gap is addressed.

Chapter 6 presents the resolution performance for multiband coherent and non-coherent systems. Optimum detectors are identified for coherent and non-coherent dualband systems, and performance bounds and behavior patterns are obtained. Novel results for the resolution performance, not necessarily better with increasing band gap, are presented.

Chapter 7 introduces spectral estimators suitable for the multiband case.

Non-coherent and novel coherent high-resolution algorithms are presented. Coherent algorithms require an exact knowledge of the band gap information, as the relation between the samples at the two subbands are explicitly used. Non-coherent algorithms, however, do not exploit the relation between data points collected at different subbands, so that the band gap information is not required.

An approach for correcting the lack of coherence between the subbands is presented in chapter 8. Using the relation between target amplitudes at the two subbands, an estimate of the coherence error can be obtained. A Monte Carlo analysis completes the chapter.

In chapter 9, an exhaustive analysis of the accuracy and resolution performance obtained with the dualband spectral estimators is shown. A qualitative behavior as predicted by theory is observed.

The last chapter summarizes this thesis and draws conclusions. Appendices with a short introduction to the *uniform theory of diffraction* (UTD) used for the signal model, and with a description of the measuring campaign complete this thesis.

2. Multiband radar signal model

For the development and testing of algorithms for multiband or ultra-wideband radar, close-to-reality models of the backscattered field of radar scenarios valid for wide bandwidths are needed. In this chapter a signal model for the radar response of an object or a group of them in a wide frequency bandwidth is presented. The scattering signal in the frequency domain is modeled by a sum of complex oscillations. Each oscillation can be interpreted as a point scatterer with constant delay and frequency dependent amplitude. This model is validated with simulated and measured data of canonical and arbitrary-shaped objects.

2.1. Radar signal model

In the common frequency bands for radar systems, the targets can be considered to be in the high-frequency range, as they usually are greater than a wavelength. Some very popular high-frequency approximations for solving scattering problems are the *geometrical optics* (GO) theory [4, 13, 30, 69] and one of its extensions, the *geometrical theory of diffraction* (GTD) [4, 8, 27, 30, 31, 41, 58, 69]. Both approximations are based on the decomposition of a scatterer in a set of individual scattering centers with a specific frequency dependence. These scattering centers correspond to specular reflections, diffraction at wedges or tips or creeping wave terms. Assuming that the high-frequency approximation is valid, the radar response of a target or a group of them can be represented as a sum of individual scattering centers:

$$s(t) = \sum_{i=1}^{P} h_i'' \delta(t - \tau_i) \tag{2.1}$$

where each Dirac's delta represents a scattering center at distance $r_i = c_0 \tau_i / 2$ with amplitude h_i''. Assuming that this model is valid along a wide frequency band, the expression for the frequency domain results in:

$$s(f) = \sum_{i=1}^{P} h_i' p_i^f \qquad \text{with } p_i = \rho_i e^{-j2\pi \tau_i} \tag{2.2}$$

where the poles p_i represent the scattering centers at distances r_i. A complex amplitude h_i' has been included in the model together with a decay of the response with frequency by means of the pole amplitude ρ_i.

This radar response is measured in the frequency domain by a system working in the frequency band $\mathcal{B}_l$, i.e. lowest frequency $\mathfrak{f}_l$, N_l measuring points and distance between them of $\delta\mathfrak{f}$. The expression of the sampled radar frequency response is

$$\mathfrak{f}_n = \mathfrak{f}_l + n\delta\mathfrak{f} \qquad\qquad \text{for } n = 0, \ldots, N_l - 1$$

$$s(\mathfrak{f}_n) = \sum_{i=1}^{P} h_i' p_i^{\mathfrak{f}_l} \cdot \left(p_i^{\delta\mathfrak{f}} \right)^n \tag{2.3}$$

$$s[n] = \sum_{i=1}^{P} h_i' z_i^{n_l} \cdot z_i^n = \sum_{i=1}^{P} h_i z_i^n$$

with $z_i = p_i^{\delta\mathfrak{f}}$, $n_l = \mathfrak{f}_l/\delta\mathfrak{f}$ and $h_i = h_i' z_i^{n_l}$.

The main assumption of this model for the radar response is that the delay of the scattering centers remains constant with frequency. This assumption is validated in the following sections by analyzing the electromagnetic scattering given by the GO and GTD theories, and analyzing the radar response of different arbitrary shaped objects obtained by means of simulations and measurements.

2.2. Scattering center analysis in GO and GTD

The GO and the GTD approximations are very intuitive theories which describe the propagation of high-frequency electromagnetic waves and their interaction with scattering objects in terms of rays. A brief introduction to both theories can be found in the appendix A, here only some results concerning the validation of the radar response model are presented.

In the high-frequency limit, the electromagnetic field can be described as propagating along trajectories called rays. The rays are everywhere orthogonal to the wavefronts in an isotropic medium and they fulfill the Fermat's principle. In homogeneous media the rays are straight lines. The field has a local plane wave nature, both electric and magnetic fields are orthogonal to the ray.

The electric field along a ray from some reference point $\vec{r}_0$ to $\vec{r}$, e.g. the ray shown in figure 2.1, has the following expression:

$$\vec{E}(\vec{r}) = \vec{E}_0(\vec{r}_0) e^{j\Phi(\vec{r}_0)} \sqrt{\frac{\rho_1\rho_2}{(\rho_1 + s)(\rho_2 + s)}} e^{-jks} \tag{2.4}$$

where ρ_1 and ρ_2 are the principal radii of curvature of the wavefront at the reference point $\vec{r}_0$, and s is the optical length from $\vec{r}_0$ to $\vec{r}$. The square root quantity is called the *divergence factor* and indicates the manner in which the energy spreads along the ray path; it is consequence of the conservation of energy. The polarization of the electric and magnetic fields does not change along the ray.

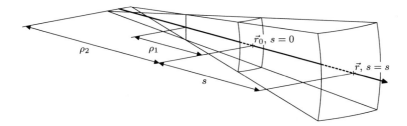

Figure 2.1: Infinitesimally narrow diverging astigmatic ray tube

The frequency dependence in the expression (2.4) for a ray traveling from $\vec{r}_0$ to $\vec{r}$ lies only on the phase term. For a given optical distance s the phase term increases linearly with the wavenumber k, resulting in a constant ray travel time.

In the GO and GTD theories, the interaction of electromagnetic field with scattering objects is also modeled in terms of rays. Depending on the surface shape, the following scattered rays are identified:

- reflected and refracted rays at smooth surfaces,

- diffracted rays at surface discontinuities like wedges or tips, and

- "creeping" rays propagating along smooth surfaces.

One of the principles of GO and GTD is the local nature of the high-frequency effects, i. e. the properties of the reflected, diffracted and creeping rays depend only on the incident field and on the surrounding area at the specular points, the diffraction points or the creeping ray trajectories. This allows to subdivide a complex scatterer in a sum of individual scattering centers, like specular points, wedges or tips. Interaction between objects is also modeled in terms of rays, a scattered ray on one object may be incident on other one and after a second scattering effect may travel to the observer. This effect will contribute as an effective scattering center at a distance which does not correspond to a point on the object.

The general expression for a scattered ray is:

$$\vec{E}(\vec{r}) = \vec{E}^i(Q) R \sqrt{\frac{\rho_1 \rho_2}{(\rho_1 + s)(\rho_2 + s)}} e^{-jks} \tag{2.5}$$

where $\vec{E}^i(Q)$ is the incident field on the scattering point Q and R is the reflection, transmission or the diffraction coefficient.

Applying the boundary conditions to a GO ray incident on a smooth surface allows to obtain the reflection/transmission coefficient R and the radii of curvature ρ_i. In GO, the scattering field is expressed in terms of direct, reflected

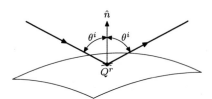

Figure 2.2: Reflected ray on a smooth surface

and refracted rays. GO fails to predict the electromagnetic field at shadow regions, i. e. where no ray is present, and therefore, the predicted field is zero. The aim of the GTD is to overcome this problem by adding diffracted rays, which are also present in shadow regions. The diffraction coefficients included in the GTD are obtained by analyzing the so-called canonical problems, i. e. scattering problems with simple geometry for which an analytical solution is available. The expression for the scattered field is rewritten in terms of GO rays to identify diffraction coefficients.

Next, some canonical scattering problems for perfectly conducting materials are analyzed.

Specular reflection

The electric field along a reflected ray can be expressed as:

$$\vec{E}^r(\vec{r}) = \vec{E}^i(Q^r)R_{s,h}\sqrt{\frac{\rho_1^r\rho_2^r}{(\rho_1^r+s)(\rho_2^r+s)}}e^{-jks^r} \tag{2.6}$$

where $\vec{E}^i(Q^r)$ is the incident field on the specular point Q^r, ρ_i^r are the radii of curvature, s^r is the path length along the reflected ray, and $R_{s,h}$ is the reflection coefficient. The reflected ray fulfills the Snell's law, i. e. the incident and the reflected angles are equal.

The frequency dependence of the phase term predicts a constant travel time from the specular point to the observer. The expressions for the radii of curvature depend only on geometrical considerations, and the reflection coefficient depends only on the polarization:

$$R_{s,h} = \mp 1 \tag{2.7}$$

where s and h stand for *soft* and *hard* boundary conditions, i. e. for electric field normal to the plane of incidence (plane formed by the incident and reflected rays) or contained on it, respectively.

The reflected ray predicted by the GO validates the assumption of the radar response model, i. e. frequency independent delay of the scattering center. The ray amplitude has also no frequency dependence.

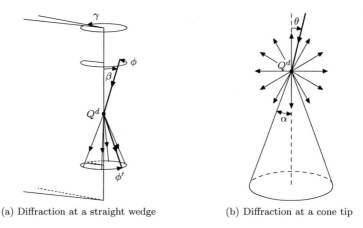

(a) Diffraction at a straight wedge (b) Diffraction at a cone tip

Figure 2.3: Diffracted rays at surface discontinuities

Diffraction at straight wedges

The general case of a ray diffracted at a straight wedge is shown in figure 2.3a. A ray is incident on the wedge and a set of infinite diffracted rays forming the same angle β with the wedge as the incident ray starts at the diffraction point Q^d.

The field along one of the diffracted rays at the wedge can be expressed as a cylindrical wave with origin at the wedge:

$$\vec{E}^d(\vec{r}) = \vec{E}^i(Q^d)D_{s,h}\frac{e^{-jks^d}}{\sqrt{s^d}} \tag{2.8}$$

where Q^d is the diffraction point, s^d the ray length starting at Q^d and $D_{s,h}$ the diffraction coefficient:

$$D_{s,h} = \frac{e^{-j\pi/4}\sin(\pi/n)}{n\sqrt{2\pi k}\sin\beta}\left(\frac{1}{\cos\left(\frac{\pi}{n}\right) - \cos\left(\frac{\phi-\phi'}{n}\right)} \pm \frac{1}{\cos\left(\frac{\pi}{n}\right) - \cos\left(\frac{\phi+\phi'}{n}\right)}\right) \tag{2.9}$$

with

$$n = \frac{\pi}{\gamma} \tag{2.10}$$

and β, ϕ and ϕ' defined as in the figure 2.3a.

The field phase term e^{-jks^d} accounts for the constant ray travel time from the diffraction point to the observer. The diffraction coefficient $D_{s,h}$ presents

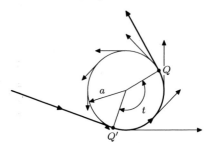

Figure 2.4: Creeping rays in a perfectly conducting infinitely long cylinder

a frequency independent phase and an absolute value decay with $\sqrt{\mathfrak{f}}$. This ray term validates also the assumption for the radar response model.

Diffraction at cone tips

The general case of a ray diffracted at a cone tip is shown in figure 2.3b. A ray is incident on the tip and a set of infinite diffracted rays ray starts at the diffraction point Q^d. For a narrow angle cone with half angle $\alpha \ll \pi/2$, angle of incidence $0 < \theta \leq \alpha$, and soft boundary condition the diffracted ray has the expression of a spherical wave with origin at the cone tip:

$$\vec{E}^d(\vec{r}) = \vec{E}^i(Q^d)D_s\frac{e^{-jks^d}}{s^d} \tag{2.11}$$

with Q^d the cone tip, s^d the ray optical length from Q^d to the observer and D_s the diffraction coefficient with the expression:

$$D_s = \frac{-j}{k} \cdot \frac{3 + \cos^2\theta}{4\cos^2\theta}\left(\frac{\alpha}{2}\right)^2 \quad \text{for} \quad \alpha \ll \pi/2 \quad \text{and} \quad 0 < \theta \leq \alpha. \tag{2.12}$$

Again, the field phase term e^{-jks^d} accounts for the constant ray travel time from the diffraction point to the observer. The diffraction coefficient D_s presents a frequency independent phase and an absolute value decay with increasing frequency $\mathfrak{f}$. This ray term validates also the assumption for the radar response model.

Creeping waves

A creeping ray is a *surface-diffracted* ray which propagates along a smooth surface transporting energy to the shadow zone, also in the case where no sharp edges or tips are present. A creeping ray becomes "attached" to a smooth surface at points of grazing incidence, at which it is tangential to the surface, travels along the geodesic path of the object, and leaves the object tangentially. This creeping ray spreads its energy in all the directions while propagating along the surface.

For the two dimensional case of an infinitely long perfectly conducting cylinder, following the development from Keller, the creeping ray has the expression of a cylindrical wave:

$$\vec{E}^d(\vec{r}) = \vec{E}^i(Q')T_{s,h}\frac{e^{-jks^d}}{\sqrt{s^d}} \tag{2.13}$$

where $\vec{E}^i(Q')$ is the GO type incident field at the attachment point Q', s^d is the optical path from the shedding point Q to the observer and $T_{s,h}$ is the surface diffraction coefficient, which can be expressed as:

$$T_{s,h} = \left(\sum_{n=1}^{N}(D_n^{s,h})^2 e^{-\alpha_n^{s,h}t}\right)e^{-jkt} \tag{2.14}$$

with

$$(D_n^s)^2 = \sqrt{\frac{1}{2\pi k}}\left(\frac{ka}{2}\right)^{\frac{1}{3}}\frac{e^{-j\pi/12}}{[\text{Ai}'(-q_n)]^2}$$
$$(D_n^h)^2 = \sqrt{\frac{1}{2\pi k}}\left(\frac{ka}{2}\right)^{\frac{1}{3}}\frac{e^{-j\pi/12}}{q_n'[\text{Ai}(-q_n')]^2} \tag{2.15}$$

$$\alpha_n^s = \frac{q_n}{a}\left(\frac{ka}{2}\right)^{\frac{1}{3}}e^{j\pi/6}$$
$$\alpha_n^h = \frac{q_n'}{a}\left(\frac{ka}{2}\right)^{\frac{1}{3}}e^{j\pi/6} \tag{2.16}$$

where t is the cylinder length traveled by the creeping ray. The quantities $(-q_n)$ are the the zeros of the Airy function Ai, $\text{Ai}(-q_n) = 0$, and $(-q_n')$ the zeros of the derivative of the Airy function Ai', $\text{Ai}'(-q_n') = 0$.

As in the previous cases, the field phase term e^{-jks^d} accounts for the constant ray travel time from the shedding point to the observer. The frequency dependence of the surface diffraction coefficient phase term is mainly given by e^{-jkt}, which establishes a constant travel time along the cylinder surface. The absolute value presents a decay with increasing frequency. The constant travel time of the ray along the whole path validates the assumptions made for the radar response model.

Analysis of some canonical problems

It has been seen that the ray terms in the GO and GTD theories validate the assumptions made for the radar signal model, i. e. the scattering of arbitrarily shaped objects can be subdivided into contributions of individual scattering centers which have a constant delay and a frequency dependent amplitude. Figures 2.5, 2.6 and 2.7 show the scattering behavior of reflected and diffracted

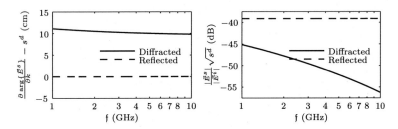

Figure 2.5: Reflected and surface diffracted rays on an infinite long cylinder with radius $a = 3$ cm obtained with GTD. Two dimensional case, plane wave incidence, $\theta_i = 0°$, $t = 180°$ and hard boundary condition

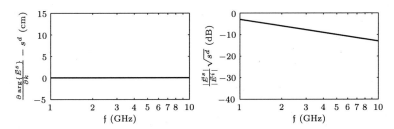

Figure 2.6: Diffracted ray on the wedge of an infinite long wedge with $\gamma = 330°$ obtained with GTD. Plane wave incidence with $\beta = 90°$, $\phi = \phi' = 120°$ and hard boundary condition

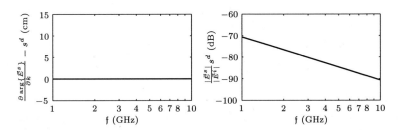

Figure 2.7: Diffracted ray on the tip of an infinite long cone with $\alpha = 9°$ obtained with GTD. Plane wave incidence with $\theta = 0°$

rays for some canonical problems. It is seen that the scattering center position of each ray does not vary with frequency. Only in the creeping wave term, a small variation with frequency of $1\,\mathrm{cm}/\mathrm{decade}$ is observed. The ray amplitudes present a decay with increasing frequency. The maximum decay is observed for the cone tip diffraction, with $20\,\mathrm{dB}/\mathrm{octave}$, which correspond with the $1/f$ amplitude dependence.

2.3. Scattering center analysis of simulated and measured objects

To validate the radar signal model, the backscattering responses of different objects, available for a wide frequency bandwidth, are analyzed to obtain the signal poles which best fit the data at different frequency subbands.

The scattering data is split into subbands of approx. $2\,\mathrm{GHz}$, 64 samples with a sampling distance between them of $\delta f = 31.5\,\mathrm{MHz}$, which results in a Rayleigh resolution distance of $7.5\,\mathrm{cm}$. Each data set is analyzed with the *modified covariance* Root-MUSIC algorithm, as described in chapter 3, where the data window length is set to $L = 30$, and the number of poles P varies depending on the scenario from 1 to 3. These poles are then modified in an optimization process to reduce the energy of the difference signal between the data set and the data model.

Data sets from canonical and complex objects are used for the analysis. Their response is obtained analytically, with simulations or with radar measurements. In the measuring campaign—described in appendix B—the radar response of different objects placed in an anechoic chamber is measured in the frequency domain with a network analyzer. As simulation method the *uniform theory of diffraction* (UTD)—as described in appendix A—is used. Analysis of canonical objects give valuable results as they are geometrical simple elements of more complex objects, i.e. a general scatterer can be considered to be built with different curved surfaces, wedges, tips,..., so that the observed effects on them can be extended to more general cases. Complex objects give more confidence to the model validation.

The analyzed data sets are the backscattering response of

- a perfectly conducting sphere with radius $5\,\mathrm{cm}$ obtained analytically [69],

- two-dimensional scenario with three cylinders obtained with UTD (described with more detail in appendix A),

- scenario with three long cylinders with the same position as in the previous case but obtained in a measurement campaign in an anechoic chamber and

- a bicycle.

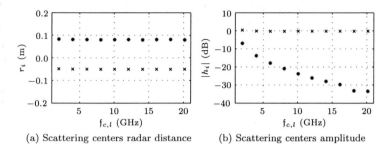

(a) Scattering centers radar distance   (b) Scattering centers amplitude

Figure 2.8: Scattering center analysis with 2 GHz data sets with center frequencies $\mathfrak{f}_{c,l}$ of a metallic sphere with radius 5 cm. Analytical solution [69] for the backscattered field

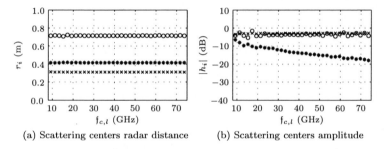

(a) Scattering centers radar distance   (b) Scattering centers amplitude

Figure 2.9: Scattering center analysis with 2 GHz data sets with center frequencies $\mathfrak{f}_{c,l}$ of a scenario with three cylinders as shown in figure A.6. Backscattered field obtained with UTD

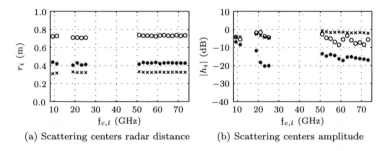

(a) Scattering centers radar distance   (b) Scattering centers amplitude

Figure 2.10: Scattering center analysis with 2 GHz data sets with center frequencies $\mathfrak{f}_{c,l}$ of a scenario with three cylinders as shown in figure A.6. Backscattered field measured in an anechoic chamber

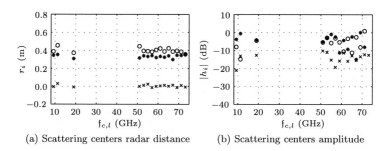

(a) Scattering centers radar distance   (b) Scattering centers amplitude

Figure 2.11: Scattering center analysis with 2 GHz data sets with center frequencies $f_{c,l}$ of a bicycle. Backscattered field measured in an anechoic chamber

The results are shown in figures 2.8, 2.9, 2.10 and 2.11. It can be seen the radar distance of each pole together with their associated amplitude as function of the subband. In the measured scenarios, the power of the signal for each measuring frequency band, i.e. X, K and V, was normalized to one, hence, a comparison is only meaningful between amplitudes of the same band. The scatterer radar distance remains constant along the different subbands, which is the main assumption of the presented radar model. The associated amplitude shows a decay for some centers, for the creeping wave term from the sphere or for the second cylinder, which is behind the first one and with increasing frequency the shadowing effect becomes larger. The maximum decay is 25 $^{dB}/_{decade}$, corresponding to the creeping ray term in the sphere. For the measured objects, the interpretation of the amplitude is more difficult.

2.4. Conclusions

In this chapter a radar signal model valid for the high-frequency range is presented. A general radar response can be decomposed into contributions of individual scattering centers which have a constant radar distance and a frequency dependent amplitude. The GO and GTD theories validate this model, predicting a constant radar distance and a maximum amplitude decay of 20 $^{dB}/_{decade}$. Analysis of simulated and measured radar responses of canonical and complex objects are also presented. The radar distances remain constant, and a decay in the amplitudes with increasing frequency can be observed.

The scattering amplitude decay is modeled by a signal pole p_i which does not lie on the unit circle, i.e. its absolute value is lower than one. As example consider a scattering center with a decay of 20 $^{dB}/_{decade}$, which is evaluated along a bandwidth of from 10 GHz to 14 GHz and sampled each 35 MHz. The absolute value of the pole which best fits the given response is 0.9971, very close to 1. Hence, it is meaningful for further analysis to consider the signal poles p_i laying on the unit circle.

3. Spectral estimation for single-band radar systems

The principal function of a radar system is to detect and range scatterers in its environment by means of electromagnetic waves. This scattering information can be represented by the *range profile*, i. e. the amplitude of the reflected wave as a function of the distance to the radar. This range profile can be measured directly in time domain—as a pulse radar does—or in frequency domain—as e. g. in a *stepped frequency continuous wave* (SFCW) radar—as both domains are related to each other by the Fourier transform. The bandwidth used by the radar is always restricted—as the duration of a pulse can not be infinitesimally small or any measurement system cover frequencies from zero to infinity—and therefore, the range profile must be always estimated using only a limited interval of the spectral information. Due to the symmetry of the Fourier transform, techniques to estimate the spectral density from signals observed during a limited period of time, the so-called *spectral estimation* techniques, can be used for the range profile estimation from radar signals in the frequency domain with a limited bandwidth. It is observed from the radar signal model presented in chapter 2 that each scattering center results in a complex oscillation in the frequency domain. Detection and ranging of scattering centers in the frequency domain is equivalent to detection and frequency estimation of complex oscillations in the time domain, which is a special technique of spectral estimation algorithms.

There are several spectral estimation techniques (see e. g. [57]) which are nowadays still object of analysis and investigation. Spectral estimation techniques are usually classified in *classical, parametric* and *eigenvalue-based* techniques. The classical spectral estimation techniques are based on the *discrete Fourier transform* (DFT) of autocorrelation estimates. Parametric spectral estimates assumes that the available data samples fit a signal model described by a finite set of parameters. If the parameters are found, the signal would be completely characterized, and the spectral density directly obtained. The set of parameters are adapted to fit best the data, e. g. in the minimum square error sense. Eigenvalue-based techniques use the properties of the eigenvalue decomposition of the autocorrelation signal to obtain orthogonal vector spaces describing the noise and the signal independently. From this spaces, information of the exponentials contained in the signal can be extracted.

A short review of these spectral estimation and frequency estimation techniques is given in the following sections, more information can be found in the literature [28, 36, 39, 57, 65].

3.1. Classical spectral estimation

Correlogram

The classical spectral estimation is based on the use of the Wiener-Khinchin theorem, which relates the *power spectral density* (PSD) of a stationary random process with its *autocorrelation sequence* (ACS)

$$P_{xx}(f) = \mathcal{F}\{r_{xx}[m]\} = \sum_{m=-\infty}^{\infty} r_{xx}[m]e^{-j2\pi fm}. \tag{3.1}$$

In practice, the ACS of a process is unknown, and only N samples of a given realization are available. The *correlogram* method substitutes a finite sequence of autocorrelation estimates for the infinite sequence of unknown true autocorrelation values

$$\hat{P}_{xx}(f) = \sum_{m=-L}^{L} \hat{r}_{xx}[m]e^{-j2\pi fm}. \tag{3.2}$$

For $\hat{r}_{xx}[m]$ the unbiased ACS estimate can be used

$$\hat{r}_{xx}[m] = \frac{1}{N-m} \sum_{n=0}^{N-m-1} x[n+m]x^*[n]. \tag{3.3}$$

Although $\hat{r}_{xx}[m]$ is unbiased, the correlogram method results in a biased estimation of the PSD. The use of a finite number of autocorrelation lags produces an estimate that is effectively the convolution of the true PSD with the transform of the rectangular window of length L, i. e. the sinc-function, which has a well-known 3 dB-width of $0.89/L$ and a -13 dB sidelobe level. Narrowband signals or lines in the spectrum will show a width of $0.89/L$ in the PSD estimate, and even at frequencies where the true PSD is zero, a nonzero PSD is estimated due to the sidelobes of the sinc-function. This sidelobe energy is called the *leakage*. Through the parameter L, a trade-off between resolution and variance is established. A high value for L implies a narrower sinc-function, and therefore a better *resolution*, i. e. the capability to detect two closely spaced spectral lines as two, but increases also the variance of the estimate, as for the high autocorrelation lags, a lower number of data samples can be used. A maximum of $L \approx N/10$ was suggested by Blackman and Tukey as optimum.

The most general form of the correlogram method of PSD estimation is called the *Blackman-Tukey* approach [7] and takes the form

$$P_{\mathrm{BT}}(f) = \sum_{m=-L}^{L} w[m]\hat{r}_{xx}[m]e^{-j2\pi fm} \tag{3.4}$$

where $w[m]$ is a general window of length $(2L + 1)$. Through the selection of the window different resolution and leakage characteristics can be obtained.

Periodogram

An alternative definition of the PSD for ergodic processes is

$$P_{xx}(f) = \lim_{N \to \infty} E \left\{ \frac{1}{2N+1} \left| \sum_{n=-N}^{N} x[n] e^{-j2\pi fn} \right|^2 \right\}. \tag{3.5}$$

Ignoring the expectation and assuming a finite dataset of N samples, the sample spectrum

$$\tilde{P}_{xx}(f) = \frac{1}{N} \left| \sum_{n=0}^{N-1} x[n] e^{-j2\pi fn} \right|^2 \tag{3.6}$$

may be computed from the finite data sequence. This is the original *unmodified periodogram* PSD estimate. The sample spectrum will yield statistically inconsistent PSD estimates because the expectation operator in eq. (3.5) has been ignored. To smooth the periodogram PSD estimate, three types of basic averaging schemes have been considered.

The method of *Daniell* [15] smooths the periodogram by averaging over adjacent frequency bins, i.e. a low-pass filter is applied to the periodogram. This reduces the variance and also the resolution.

The method of *Bartlett* [6] averages multiple periodograms produced from segments of length L of the original data sequence. By averaging $M = N/L$ estimates, as the noise subsequences are considered independent to each other, the variance is reduced by a factor M. Due to the shorter data segments for the periodograms, the resolution is also reduced by a factor of M.

The method of *Welch* [94] extends the Bartlett approach by overlapping the segments and introducing data windows to improve the bias caused by leakage. The variance of the Welch method is smaller than in the Bartlett method, as the increased number of single periodograms is predominant to the loss of independence of the noise sub-sequences.

Further information and analysis of the classical spectral estimates, and combination of correlogram and periodogram techniques can be found in the comprehensive literature over classical spectral estimation [36, 39, 57, 65].

3.2. Autoregressive spectral estimation

A wide extended method to model sequences is to suppose them as result of filtering white noise. If the generator filter is found, the spectral characteristic of the sequence is coincident with that of the filter, as the white noise has a constant spectral density.

Depending on the structure of the generator filter—assuming a rational power spectra—the resulting sequence can be classified as a *moving average* (MA) signal, for *all zero* filters, an *autoregressive* (AR) signal for *all pole* filters, and as an *autoregressive-moving average* (ARMA) signal, for generator filters with poles and zeros. The MA model is suitable for representing spectra with notches, the AR model for spectra with narrow peaks, and the ARMA model for spectra with peaks and notches. A radar range profile shows sharp peaks at the scatterer positions, so that the AR model will best fit to the radar problem. Besides, the estimates of the AR parameters can be obtained as solutions to linear equations.

Consider an AR generator filter of order L.

$$H_{\mathrm{AR}}(z) = \frac{1}{A(z)} = \frac{1}{1 + a_1 z^{-1} + a_2 z^{-2} + \cdots + a_L z^{-L}} \tag{3.7}$$

The coefficients of the polynomial $A(z)$ relate the *complex white Gaussian noise* (CWGN) input sequence, $w[n] \sim \mathcal{CN}(0, \sigma_w^2)$, with the output sequence $x[n]$

$$x[n] = w[n] - \sum_{l=1}^{L} a_l x[n-l]. \tag{3.8}$$

Using these difference equations, a set of relations between the ACS of the output signal $r_{xx}[m]$ and the filter parameters is obtained, the so-called *Yule-Walker* equations

$$r_{xx}[m] = \begin{cases} -\sum\limits_{l=1}^{L} a_l r_{xx}[m-l] & \text{for } m > 0 \\ -\sum\limits_{l=1}^{L} a_l r_{xx}[-l] + \sigma_w^2 & \text{for } m = 0 \\ r_{xx}^*[-m] & \text{for } m < 0 \end{cases} \tag{3.9}$$

which in matrix form are expressed as

$$\begin{pmatrix} r_{xx}[0] & r_{xx}[-1] & \cdots & r_{xx}[-L] \\ r_{xx}[1] & r_{xx}[0] & \cdots & r_{xx}[-L+1] \\ \vdots & \vdots & \ddots & \vdots \\ r_{xx}[L] & r_{xx}[L-1] & \cdots & r_{xx}[0] \end{pmatrix} \begin{pmatrix} 1 \\ a_1 \\ \vdots \\ a_L \end{pmatrix} = \begin{pmatrix} \sigma_w^2 \\ 0 \\ \vdots \\ 0 \end{pmatrix}. \tag{3.10}$$

The generator filter can also be defined as the inverse of the *whitening* filter, a filter which converts the input sequence back into white noise. This inverse filter is strongly related to linear prediction analysis. Consider the *forward* linear prediction estimate

$$\hat{x}_f[n] = -\sum_{l=1}^{L} a_l x[n-l] \tag{3.11}$$

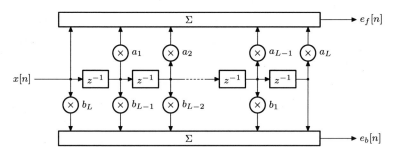

Figure 3.1: Linear prediction error filter. If $x[n]$ is an AR process of order L, the filter coefficients which minimize the error variances are identical to the AR parameters, and $e_f[n]$ and $e_b[n]$ are white processes

of the sample $x[n]$, where a_l is the filter coefficient at index l of the forward linear prediction *Finite Impulse Response* (FIR) filter of length L. The forward linear prediction error is defined as

$$e_f[n] = x[n] - \hat{x}_f[n] = x[n] + \sum_{l=1}^{L} a_l x[n-l] \tag{3.12}$$

and imposing a minimization of its variance σ_f^2 the following equations are obtained

$$\begin{pmatrix} r_{xx}[0] & r_{xx}[-1] & \cdots & r_{xx}[-L] \\ r_{xx}[1] & r_{xx}[0] & \cdots & r_{xx}[-L+1] \\ \vdots & \vdots & \ddots & \vdots \\ r_{xx}[L] & r_{xx}[L-1] & \cdots & r_{xx}[0] \end{pmatrix} \begin{pmatrix} 1 \\ a_1 \\ \vdots \\ a_L \end{pmatrix} = \begin{pmatrix} \sigma_f^2 \\ 0 \\ \vdots \\ 0 \end{pmatrix}. \tag{3.13}$$

This matrix expression has the same structure as the Yule-Walker equations in eq. (3.10). For the special case of $x[n]$ being an AR process of order L, the prediction error sequence $e_f[n]$ is a white sequence and the linear prediction coefficients are identical to the AR parameters.

Analogously, a *backward* linear prediction estimate

$$\hat{x}_b[n] = -\sum_{l=1}^{L} b_l x[n+l] \tag{3.14}$$

of the sample $x[n]$ is defined, in which b_l is the backward linear prediction coefficient at index l. The backward linear prediction error is

$$e_b[n] = x[n] - \hat{x}_b[n] = x[n] + \sum_{l=i}^{L} b_l x[n+l] \tag{3.15}$$

and imposing the minimization of its variance results in

$$b_l = a_l^*$$ (3.16)

and in identical error variance as in the forward prediction case. Again, for $x[n]$ AR process of order L, $e_b[n]$ is a white noise sequence and the backward linear prediction error filter is a whitening filter for the anti-causal realization of the AR process.

The Yule-Walker equations can be used to obtain the AR parameters for the case that L lags of the ACS are exactly known. In the general case, however, the ACS is unknown and the AR model has to be estimated from a finite number of data samples. In the Yule-Walker method, eq. (3.10) is used with an ACS estimate to obtain the AR parameters. Usually the biased form of the estimate is used to ensure stable filters.

$$\check{r}_{xx}[m] = \frac{1}{N} \sum_{n=0}^{N-|m|} x^*[n]x[n+m]$$ (3.17)

Given a finite number of samples of the signal $x[n]$, $n = 0, \ldots, N-1$, a linear system of difference equations of the forward prediction error filter can be expressed in matrix form as

$$\mathbf{e}_f = \mathbf{X}_L \begin{pmatrix} 1 \\ \mathbf{a} \end{pmatrix}$$ (3.18)

with

$$\mathbf{e}_f = \begin{pmatrix} e_f[L] \\ \vdots \\ e_f[N-1] \end{pmatrix} \quad \mathbf{a} = \begin{pmatrix} a_1 \\ \vdots \\ a_L \end{pmatrix} \quad \mathbf{X}_L = \begin{pmatrix} x[L] & \cdots & x[0] \\ \vdots & \ddots & \vdots \\ x[N-1] & \cdots & x[N-L-1] \end{pmatrix}$$

where $\mathbf{X}_L$ is called the data matrix. Similarly, the difference equations for the backward linear prediction error can be written as

$$\mathbf{e}_b^* = \mathbf{X}_L^* \mathbf{J} \begin{pmatrix} 1 \\ \mathbf{a} \end{pmatrix}$$ (3.19)

where the relation in eq. (3.16) has been included and the matrix $\mathbf{J}$ is an $(p+1) \times (p+1)$ reflection matrix, so that

$$\mathbf{e}_b = \begin{pmatrix} e_b[0] \\ \vdots \\ e_b[N-L-1] \end{pmatrix} \quad \mathbf{X}_L^* \mathbf{J} = \begin{pmatrix} x^*[0] & \cdots & x^*[L] \\ \vdots & \ddots & \vdots \\ x^*[N-L-1] & \cdots & x^*[N-1] \end{pmatrix}.$$

Writing the previous equations together results in

$$\mathbf{e} = \begin{pmatrix} \mathbf{e}_f \\ \mathbf{e}_b^* \end{pmatrix} = \begin{pmatrix} \mathbf{X}_L \\ \mathbf{X}_L^* \mathbf{J} \end{pmatrix} \begin{pmatrix} 1 \\ \mathbf{a} \end{pmatrix}.$$ (3.20)

To obtain an optimum predictor filter, in the sense of minimum square prediction error, the average of the forward and the backward linear prediction squared errors $\mathbf{e}^H\mathbf{e}$ is minimized over the available data with respect to the filter coefficients a_l ($\frac{\partial \mathbf{e}^H\mathbf{e}}{\partial a^*} = 0$) and the following expression for $\mathbf{a}$ is obtained:

$$\mathbf{a} = -\left(\mathbf{X}^H\mathbf{X}\right)^{-1}\mathbf{X}^H\mathbf{x} \tag{3.21}$$

where $\mathbf{X}$ is built with the columns 2 to $L+1$ of matrix $\begin{pmatrix} \mathbf{X}_L \\ \mathbf{X}_L^*\mathbf{J} \end{pmatrix}$:

$$\mathbf{X} = \begin{pmatrix} x[L-1] & \cdots & x[0] \\ \vdots & \ddots & \vdots \\ x[N-2] & \cdots & x[N-L-1] \\ x^*[1] & \cdots & x^*[L] \\ \vdots & \ddots & \vdots \\ x^*[N-L] & \cdots & x^*[N-1] \end{pmatrix}$$

and the vector $\mathbf{x}$ is the first column of matrix $\begin{pmatrix} \mathbf{X}_L \\ \mathbf{X}_L^*\mathbf{J} \end{pmatrix}$:

$$\mathbf{x} = \begin{pmatrix} x[L] & \cdots & x[N-1] & x^*[0] & \cdots & x^*[N-L-1] \end{pmatrix}^T.$$

This method is termed the *modified covariance* or the *unwindowed data least-squares* method [10, 60, 92]. The method which only uses the forward or backward linear prediction equations is called the *covariance* method. Another widely used method is the *autocorrelation* or *windowed* method, which instead of using the data matrix in eq. (3.18) uses an extended data matrix

$$\mathbf{X}_L' = \begin{pmatrix} x[0] & \cdots & 0 \\ \vdots & \ddots & \vdots \\ x[L] & \cdots & x[0] \\ \vdots & \ddots & \vdots \\ x[N-1] & \cdots & x[N-L-1] \\ \vdots & \ddots & \vdots \\ 0 & \cdots & x[N-1] \end{pmatrix}. \tag{3.22}$$

The modified covariance method shows the best performance in terms of resolution capability, frequency bias estimation and line splitting [57, 65]. The main drawback of both covariance methods is that they do not guarantee a stable linear prediction filter, as both autocorrelation methods do. However, this is not a problem in the spectral estimation case, as the filter parameters are used only for spectral estimation purposes.

Another widely used AR algorithm is the *Burg* or the *Maximum entropy* method [10, 11], which ensures filter stability by imposing the satisfaction of the Levinson-Durbin algorithm. Due to the fact that filter stability is not an issue in spectrum estimation, this method is not further discussed here.

Once the predictor filter coefficients are available, the spectrum of the generator filter (an estimate of the signal's spectrum) can be directly obtained using

$$P_{\text{AR}}(f) = \sigma_e^2 \cdot |H_{\text{AR}}(e^{j2\pi f})|^2 = \frac{\sigma_e^2}{|A(e^{j2\pi f})|^2} \tag{3.23}$$

with

$$A(z) = 1 + a_1 z^{-1} + \cdots + a_L z^{-L}$$

and σ_e^2 the variance of the error. If the main interest is not focused on the spectral density but on the location of the peaks, or targets for the radar range profile case, they are directly given by the poles of the generation filter, i.e. the zeros of the polynomial $A(z)$.

Order selection criteria

The quality of the spectral estimate depends also on the selected filter order L. Too low order results in highly smoothed spectral estimate and too high order increases the resolution but introduces spurious detail into the spectrum. The classical trade-off between resolution and estimate variance must be established. A limit to the order is given by the following necessary, but not sufficient, condition for the data matrices to be non-singular [57], i.e. $L \leq N/2$ for

$$\mathbf{X}_L^H \mathbf{X}_L$$

and $L \leq 2N/3$ for

$$\begin{pmatrix} \mathbf{X}_L \\ \mathbf{X}_L^* \mathbf{J} \end{pmatrix}^H \begin{pmatrix} \mathbf{X}_L \\ \mathbf{X}_L^* \mathbf{J} \end{pmatrix} .$$

Several methods to estimate the proper model order are proposed in the literature, e.g. *final prediction error* [2], the *Akaike information criterion* (AIC) [3], the *minimum description length* (MDL) [67] or the *criterion autoregressive transfer* (CAT) [62]. This thesis does not deal with this problem, as it is assumed that the proper model order is known, so that it will be not further discussed.

3.3. Eigenanalysis based frequency estimation

The eigenanalysis-based algorithms use the division of the information in the autocorrelation matrix into two orthogonal vector subspaces: a *signal plus noise* subspace, and a *noise* subspace. The orthogonality between these two vector subspaces is used to create *frequency estimators* of the sinusoids or narrowband spectral components present in the signal. These algorithms are not true PSD estimators because they do not preserve the measured process power, nor can the autocorrelation sequence be recovered by Fourier transforming the frequency estimators.

Next, the *multiple signal classification* (MUSIC) [71, 72] algorithm is presented. Other eigenvalue-based popular methods, as the *Pisarenko harmonic decomposition* (PHD) [64] or the *estimation of signal parameters via rotational invariance techniques* (ESPRIT) [68] are beyond the scope of this thesis, as the dualband analysis is best shown in the MUSIC algorithm.

Consider the complex noiseless sequence $s[n]$ composed by a sum of P *complex damped oscillations* which can be expressed as:

$$s[n] = \sum_{i=1}^{P} h_i z_i^n \quad \text{for} \quad n = 0, \ldots, N-1. \tag{3.24}$$

The related order L data matrix $\mathbf{S}_L$—equal to the order L data matrix of the covariance method of linear prediction—is defined as:

$$\mathbf{S}_L = \begin{pmatrix} s[L] & \cdots & s[0] \\ \vdots & \ddots & \vdots \\ s[N-1] & \cdots & s[N-L-1] \end{pmatrix} \tag{3.25}$$

and it can be expressed as the product of two rank P matrices

$$\mathbf{S}_L = \mathbf{B} \cdot \mathbf{C} \tag{3.26}$$

in which $\mathbf{B}$, an $(N-L) \times P$ matrix, and $\mathbf{C}$, an $P \times (L+1)$ matrix, are

$$\mathbf{B} = \begin{pmatrix} h_1 z_1^L & \cdots & h_P z_P^L \\ h_1 z_1^{L+1} & \cdots & h_P z_P^{L+1} \\ \vdots & \ddots & \vdots \\ h_1 z_1^{N-1} & \cdots & h_P z_P^{N-1} \end{pmatrix} \tag{3.27}$$

$$\mathbf{C} = \begin{pmatrix} 1 & z_1^{-1} & \cdots & z_1^{-L} \\ 1 & z_2^{-1} & \cdots & z_2^{-L} \\ \vdots & \vdots & \ddots & \vdots \\ 1 & z_P^{-1} & \cdots & z_P^{-L} \end{pmatrix}. \tag{3.28}$$

To do an eigenvalue decomposition of the autocorrelation-like matrix $\mathbf{S}_L^H \mathbf{S}_L$ it is helpful to decompose first the $P \times P$ matrix $\mathbf{B}^H \mathbf{B} \mathbf{C} \mathbf{C}^H$, which is positive definite (both $\mathbf{B}$ and $\mathbf{C}$ are of full rank P) and has P eigenvalues. If λ_k^s for $1 \leq k \leq P$ designates its eigenvalues and $\mathbf{v}_k'$ its eigenvectors, then:

$$(\mathbf{B}^H \mathbf{B} \mathbf{C} \mathbf{C}^H) \mathbf{v}_k' = \lambda_k^s \mathbf{v}_k' \tag{3.29}$$

for $1 \leq k \leq P$. Premultiplying by matrix $\mathbf{C}^H$ yields to

$$\mathbf{C}^H \mathbf{B}^H \mathbf{B} \mathbf{C}(\mathbf{C}^H \mathbf{v}_k') = \lambda_k^s (\mathbf{C}^H \mathbf{v}_k') \tag{3.30}$$

and defining the vector

$$\mathbf{C}^H \mathbf{v}_k' = \mathbf{v}_k \tag{3.31}$$

leads to the result

$$\mathbf{S}_L^H \mathbf{S}_L \mathbf{v}_k = \lambda_k^s \mathbf{v}_k \tag{3.32}$$

for $1 \leq k \leq P$. The P nonzero eigenvalues of the matrix $\mathbf{S}_L^H \mathbf{S}_L$ are identical to the eigenvalues of the matrix $\mathbf{B}^H \mathbf{B} \mathbf{C} \mathbf{C}^H$, and the remaining $L + 1 - P$ eigenvalues are zero because the rank of the matrix is P. The eigenvectors are given by eq. (3.31), thus, any principal eigenvector of the matrix $\mathbf{S}_L^H \mathbf{S}_L$ is a linear combination of the columns of $\mathbf{C}^H$ which is composed by the signal vectors as shown in eq. (3.28).

Considering that the sequence $s[n]$ is measured in the presence of additive CWGN, $w[n] \sim \mathcal{CN}(0, \sigma_w^2)$, the measured signal is

$$x[n] = \sum_{i=1}^{P} h_i z_i^n + w[n] = s[n] + w[n] \qquad \text{for} \quad n = 0, \ldots, N-1. \tag{3.33}$$

The related data matrix $\mathbf{X}_L$ of order L is

$$\mathbf{X}_L = \mathbf{S}_L + \mathbf{W}_L \tag{3.34}$$

and the autocorrelation-like matrix

$$\begin{aligned} \mathbf{X}_L^H \mathbf{X}_L &= (\mathbf{S}_L^H + \mathbf{W}_L^H) \cdot (\mathbf{S}_L + \mathbf{W}_L) \\ &= \mathbf{S}_L^H \mathbf{S}_L + \mathbf{W}_L^H \mathbf{W}_L + \mathbf{S}_L^H \mathbf{W}_L + \mathbf{W}_L^H \mathbf{S}_L. \end{aligned} \tag{3.35}$$

When the number of data samples tends to infinity the time average equals the expected value and

$$\lim_{N \to \infty} (\mathbf{X}_L^H \mathbf{X}_L) = \mathbf{R}_{xx} \tag{3.36a}$$

$$\lim_{N \to \infty} (\mathbf{W}_L^H \mathbf{W}_L) = \sigma_w^2 \mathbf{I}_{L+1} \tag{3.36b}$$

$$\lim_{N \to \infty} (\mathbf{S}_L^H \mathbf{W}_L) = 0 \tag{3.36c}$$

$$\lim_{N \to \infty} (\mathbf{W}_L^H \mathbf{S}_L) = 0. \tag{3.36d}$$

Considering the number of data samples N is high enough to assume the limit values as valid, the equation (3.35) yields to

$$\mathbf{X}_L^H \mathbf{X}_L = \mathbf{S}_L^H \mathbf{S}_L + \sigma_w^2 \mathbf{I}_{L+1}. \tag{3.37}$$

Substituting $\mathbf{S}_L^H \mathbf{S}_L$ and $\sigma_w^2 \mathbf{I}_{L+1}$ by their eigenvalue decompositions

$$\begin{aligned}
\mathbf{X}_L^H \mathbf{X}_L &= \sum_{k=1}^{P} \lambda_k^s \mathbf{v}_k + \sum_{k=1}^{L+1} \sigma_w^2 \mathbf{v}_k \\
&= \sum_{k=1}^{P} (\lambda_k^s + \sigma_w^2) \mathbf{v}_k + \sum_{k=P+1}^{L+1} \sigma_w^2 \mathbf{v}_k
\end{aligned} \tag{3.38}$$

the eigenvalue decomposition of $\mathbf{X}_L^H \mathbf{X}_L$ is achieved. Its eigenvectors can be subdivided in the *signal plus noise* subspace, the principal ones of the $\mathbf{S}_L^H \mathbf{S}_L$ matrix, and the *noise* subspace. The noise subspace will be orthogonal to the signal plus noise subspace and therefore, attending to eq. (3.31) and to the fact that $\mathbf{C}^H$ is of rank P, its eigenvectors are also orthogonal to the columns of $\mathbf{C}^H$ or any combination of them,

$$\mathbf{s}^H(z_i) \sum_{k=P+1}^{L+1} \psi_k \mathbf{v}_k = 0 \tag{3.39}$$

where ψ_k are arbitrary weighting factors and $\mathbf{s}(z_i)$ are the columns of $\mathbf{C}^H$

$$\mathbf{s}(z_i) = \begin{pmatrix} 1 & (z_i^*)^{-1} & \cdots & (z_i^*)^{-L} \end{pmatrix}^T \tag{3.40}$$

so that

$$\mathbf{s}^H(z_i) \sum_{k=P+1}^{L+1} \psi_k \mathbf{v}_k = \sum_{k=P+1}^{L+1} \psi_k \left(v_k[1] + v_k[2] z_i^{-1} + \cdots + v_k[L+1] z_i^{-L} \right) = 0$$

and z_i will be roots of any combination of the *noise polynomials*

$$V_k(z) = v_k[1] + v_k[2] z^{-1} + \cdots + v_k[L+1] z^{-L} \qquad k = P+1, \ldots, L+1 \tag{3.41}$$

For a signal composed of *undamped* complex oscillations an analogous analysis can be done using the modified covariance data matrix. Because of the increased number of rows in the data matrix, a better performance is expected.

Consider the signal $s[n]$ received in the presence of additive CWGN $w[n]$ with variance σ_w^2

$$x[n] = \sum_i h_i e^{j\omega_i} + w[n] = s[n] + w[n] \qquad n = 0, \ldots, N-1. \tag{3.42}$$

The modified covariance data matrix is defined as in eq. (3.20)

$$\begin{pmatrix} \mathbf{X}_L \\ \mathbf{X}_L^* \mathbf{J} \end{pmatrix} = \begin{pmatrix} x[L] & \cdots & x[0] \\ \vdots & \ddots & \vdots \\ x[N-1] & \cdots & x[N-L-1] \\ x^*[0] & \cdots & x^*[L] \\ \vdots & \ddots & \vdots \\ x^*[N-L-1] & \cdots & x^*[N-1] \end{pmatrix}. \tag{3.43}$$

Following the same analysis as in the previous case, the modified covariance data matrix of the signal $s[n]$ can be decomposed as

$$\begin{pmatrix} \mathbf{S}_L \\ \mathbf{S}_L^* \mathbf{J} \end{pmatrix} = \mathbf{BC} \tag{3.44}$$

where

$$\mathbf{B} = \begin{pmatrix} h_1 e^{j\omega_1 L} & \cdots & h_P e^{j\omega_P L} \\ h_1 e^{j\omega_1(L+1)} & \cdots & h_P e^{j\omega_P(L+1)} \\ \vdots & \ddots & \vdots \\ h_1 e^{j\omega_1(N-1)} & \cdots & h_P e^{j\omega_P(N-1)} \\ h_1^* & \cdots & h_P^* \\ h_1^* e^{-j\omega_1} & \cdots & h_P^* e^{-j\omega_P} \\ \vdots & \ddots & \vdots \\ h_1^* e^{-j\omega_1(N-L-1)} & \cdots & h_P^* e^{-j\omega_P(N-L-1)} \end{pmatrix} \tag{3.45}$$

$$\mathbf{C} = \begin{pmatrix} 1 & e^{-j\omega_1} & \cdots & e^{-j\omega_1 L} \\ 1 & e^{-j\omega_2} & \cdots & e^{-j\omega_2 L} \\ \vdots & \vdots & \ddots & \vdots \\ 1 & e^{-j\omega_P} & \cdots & e^{-j\omega_P L} \end{pmatrix}. \tag{3.46}$$

Applying the approximations in eq. (3.36), the eigenvalue decomposition of the autocorrelation-like matrix is obtained

$$\begin{pmatrix} \mathbf{X}_L \\ \mathbf{X}_L^* \mathbf{J} \end{pmatrix}^H \begin{pmatrix} \mathbf{X}_L \\ \mathbf{X}_L^* \mathbf{J} \end{pmatrix} = \sum_{k=1}^{P} (\lambda_k^s + \sigma_w^2) \mathbf{v}_k + \sum_{k=P+1}^{L+1} \sigma_w^2 \mathbf{v}_k. \tag{3.47}$$

The eigenvectors can be subdivided again into the signal plus noise subspace ($k \leq P$) and the noise subspace ($k > P$), whose eigenvectors are orthogonal to the columns of $\mathbf{C}^H$ or any combination of them

$$\mathbf{s}^H(e^{j\omega_i}) \sum_{k=P+1}^{L+1} \psi_k \mathbf{v}_k = 0 \tag{3.48}$$

where ψ_k are arbitrary weighting factors and $\mathbf{s}(e^{j\omega_i})$ are the columns of $\mathbf{C}^H$

$$\mathbf{s}(e^{j\omega_i}) = \begin{pmatrix} 1 & e^{j\omega_i} & \cdots & e^{j\omega_i L} \end{pmatrix}^T \tag{3.49}$$

The exponential terms $e^{j\omega_i}$ will be roots of any combination of the *noise polynomials*

$$V_k(z) = v_k[1] + v_k[2]z^{-1} + \cdots + v_k[L+1]z^{-L}. \tag{3.50}$$

The following relation is also valid

$$\mathbf{s}^H(e^{j\omega_i}) \left(\sum_{k=P+1}^{L+1} \psi_k \mathbf{v}_k \mathbf{v}_k^H \right) \mathbf{s}(e^{j\omega_i}) = 0 \tag{3.51}$$

so that $e^{j\omega_i}$ will also be roots of any combination of

$$V_k(z) V_k^*(1/z^*). \tag{3.52}$$

MUSIC

The MUSIC or *Spectral* MUSIC algorithm assumes that the signal poles lie on the unit circle. The algorithm is based on the detection of the peaks of the following pseudo-spectra:

$$P_{\text{MUSIC}}(f) = \frac{1}{D(e^{j2\pi f})} = \frac{1}{\mathbf{s}^H(e^{j\omega}) \left(\sum_{k=P+1}^{L} \psi_k \mathbf{v}_k \mathbf{v}_k^H \right) \mathbf{s}(e^{j\omega})} \tag{3.53}$$

where $D(e^{j2\pi f})$ is called the *null spectra*, $\mathbf{s}(e^{j\omega})$ is the *steering* or *frequency vector*

$$\mathbf{s}(e^{j\omega}) = \begin{pmatrix} 1 & e^{j\omega 1} & e^{j\omega 2} & \cdots & e^{j\omega L} \end{pmatrix}^T \tag{3.54}$$

and ψ_k is a weighting factor. The peaks will occur at the zeros of the null spectra, at the frequencies whose steering vectors are orthogonal to those of the noise subspace, i.e. at the frequencies contained in the signal $x[n]$.

Depending on the selection of the weighting factor, different methods are identified. For $\psi_k = 1$ for all k, the method is called the MUSIC algorithm. Choosing $\psi_k = 1/\lambda_k$ yields the *eigenvector* (EV) algorithm [34]. The EV method produces fewer spurious peaks than MUSIC for a given choice of order L due to the use of inverse eigenvalue weighting [57].

Root-MUSIC

As its spectral counterpart, the *Root-MUSIC* algorithm [5] assumes that the oscillations present on the signal are not damped. Instead of evaluating the position of the peaks of the pseudo-spectra, i. e. test for orthogonality along the whole unit circle, the idea behind the Root-MUSIC method is to obtain directly the roots of the null spectra polynomial:

$$D(z) = \sum_{k=P+1}^{L} V_k(z) V_k^*(1/z^*) \tag{3.55}$$

where the weighting factors ψ_k has been set to one, and $V_k(z)$ is defined in eq. (3.50)

$$V_k(z) = v_k[1] + v_k[2] z^{-1} + \cdots + v_k[L+1] z^{-L}.$$

The roots z_j will appear together with their reflected $1/z_j^*$, and due to different errors they will not necessary lie on the unit circle. Therefore, only the frequencies (the angle information) determined by the roots on or inside the unit circle (the reflected roots will lie farther from the unit circle) are necessary to be taken into account.

For the case of damped exponentials present in the signal, a small modification is necessary in the Root-MUSIC algorithm to estimate the exact position of the poles. The covariance data matrix is used and the null spectra polynomial is defined as:

$$D(z) = \sum_{k=P+1}^{L} V_k(z) \tag{3.56}$$

where $V_k(z)$ is defined in eq. (3.50)

$$V_k(z) = \left(v_k[1] + v_k[2] z^{-1} + \cdots + v_k[L+1] z^{-L} \right).$$

Order selection criteria

As in the AR algorithms, the quality of the spectral estimate in the MUSIC depends also on the selected parameters. The AIC order-selection criterion for AR algorithms has been extended to handle the subspace separation problem, i. e. selection of P [93]. An alternative algorithm based also on the eigenvector decomposition and on matrix perturbation analysis is also found in the literature [21]. This problem is beyond the scope of this thesis and will not be further discussed.

4. Range accuracy and resolution in single-band radar systems

Range accuracy and resolution are widely used parameters to characterize the quality of radar range estimation. Range accuracy is defined as the exactitude in range estimation and can be very well characterized by the mean and the variance of the estimate. Range resolution is a more abstract parameter and is defined as the capability to detect two closely spaced objects as two independent ones. Of course, accuracy and resolution are strong related to each other, as an accurate estimation can only be achieved when the observed targets have been independently detected. However, in the radar community this two parameters have been traditionally analyzed separately.

This chapter focuses on general methods to characterize performance bounds. Analysis of concrete radar systems, e. g. pulsed or FMCW, have been subject of intensive investigation and will not be treated here. Results and discussions can be found in many radar books [12, 16, 49, 55, 59, 63, 83, 84, 96].

The accuracy and resolution capabilities of spectral estimation techniques have also been extensively reported in the literature: in more general books and articles covering different approaches [28, 36, 39, 57, 65] or in articles for specific algorithms, e. g. AR [25, 54, 56, 100] or MUSIC [17, 20, 35, 44, 45, 46, 66, 86, 87, 88, 99, 101], and will not be discussed here.

Next, two classical resolution bounds are briefly presented: the width of the sinc-function and the ambiguity function. The bound given by the sinc-function width is widely used in linear spectral estimation techniques based on the Fourier transform (see e g. [36, 57, 65]). The ambiguity function is the classical tool of the radar signal theory [52, 63, 83, 84]. Sections 4.3 and 4.4 introduce the estimation and the detection theories with their application to radar signals. Both provide useful performance bounds for accuracy and resolution, respectively.

4.1. Width of the sinc-function

The most popular resolution bound for classical spectral analysis is the envelope width of a single-oscillation estimate. Classical spectral estimation is based on the use of the *discrete Fourier transform* (DFT) over a length-limited data set. Assuming rectangular windowing, the response to a single complex oscillation with frequency f_c is a sinc-function centered at f_c

$$\mathcal{F}\left\{e^{j2\pi f_c n}\text{rect}\left(\frac{n}{N}\right)\right\} = N\frac{\sin(N\pi(f - f_c))}{N\pi(f - f_c)} \tag{4.1}$$

where

$$\text{rect}\left(\frac{n}{N}\right) = \begin{cases} 1 & |n| \leq N/2 \\ 0 & |n| > N/2 \end{cases}$$

As the DFT is a linear operation, the estimate of a signal with two oscillations is the sum of two sinc-functions. If both sinc-functions are independently identified in the sum signal, i.e. two peaks are observed, the two tones are said to be resolved. In order to observe two peaks in the sum signal, a minimum separation between the sinc-functions is necessary. This separation acts as a resolution bound. The most popular definition for minimum resolvable distance is called the *Rayleigh* distance, and is given by the separation between the maximum and the first zero of the sinc-function, i.e. $1/N$. Actually, Siart shows that a separation of 40% over the Rayleigh distance is necessary to clearly detect two peaks in the spectrum [75].

Depending on the length of the data set transformed with the DFT, different resolution bounds are achieved. Techniques with longer data sets, e.g. periodogram, will achieve better resolution as techniques with subwindows or shorter data sets, e.g. Bartlett, Welch or Blackman-Tukey.

This resolution bound depends only on the signal bandwidth (or data set length). Actually, a subliminal assumption of high enough *signal-to-noise ratio* (SNR) is assumed, as peaks in the spectrum can only be detected if they are over the noise floor. However, this method does not allow to quantify the effect of the SNR in the resolution performance.

4.2. Ambiguity function

The ambiguity function represents the time response of a filter matched to a given finite energy signal when the signal is received with a delay τ and a Doppler shift ν relative to the nominal values (zeros) expected by the filter. The matched filter response is expressed as

$$\chi(\tau, \nu) = \int\limits_{-\infty}^{\infty} s(t)s^*(t + \tau)e^{j2\pi\nu t}\, dt \tag{4.2}$$

where $s(t)$ is the complex envelop of the radar signal and $|\chi(\tau, \nu)|^2$ the ambiguity function. The ambiguity function is a major tool for studying and analyzing radar signals. Some insight into these two-dimensional functions can be obtained from their one-dimensional cuts. The cut along the delay axis, coincident with the autocorrelation function, is used to assess the resolution performance of the radar signal $s(t)$. Two closely spaced targets can be detected, if their pulses at the received signal do not overlap. As matched filtering is a linear operation, the 3 dB-width of the received signal from a single pulse, i. e. the 3 dB-width of the autocorrelation signal, can be used as resolution bound. Analogously, the cut along the Doppler axis, gives a bound for the speed resolution.

Using the Taylor series expansion of $|\chi(\tau,\nu)|^2$ around the origin, a measure for the delay extent $\Delta\tau$, i. e. of the minimum resolvable distance, can be obtained

$$\Delta\tau = \frac{1}{\beta} \tag{4.3}$$

where β is the *effective bandwidth* defined by Gabor [22]

$$\beta^2 = \frac{\int_{-\infty}^{\infty} (2\pi f)^2 \, |S(f)|^2 \, df}{\int_{-\infty}^{\infty} |S(f)|^2 \, df}. \tag{4.4}$$

As seen in detection theory (section 4.4), the energy of the difference between two signals gives a measure of the capability to discriminate them. The energy of the difference between two received radar signals with different delays increase with decreasing ambiguity function, so that resolution between the two pulses becomes more probable with lower $|\chi(\tau,\nu)|^2$, result which is consistent with the autocorrelation width criteria.

Another bound for resolution performance involving the ambiguity function was given by Woodward [96]. He defined the *time resolution constant*, for use in situations involving no Doppler shift

$$T_R(0) = \frac{\int_{-\infty}^{\infty} |\chi(\tau,0)|^2 \, d\tau}{|\chi(0,0)|^2} = \frac{\int_{-\infty}^{\infty} |S(f)|^4 \, df}{\left(\int_{-\infty}^{\infty} |S(f)|^2 \, df\right)^2}. \tag{4.5}$$

The waveform which furnished the smallest value for $T_R(0)$ was assumed to have the greatest potential for resolving between two signals in time. It is seen that $T_R(0)$ is related to the reciprocal of the occupied bandwidth of a waveform. Suppose a multiband signal in the frequency domain consisting of L rectangular subbands of width B_l centered at f_l

$$S(f) = \begin{cases} U_0 & |f - f_l| \leq \frac{B_l}{2} \\ 0 & \text{elsewhere} \end{cases} \tag{4.6}$$

its time resolution constant is

$$T_R(0) = \frac{1}{\sum_{l=1}^{L} B_l} \tag{4.7}$$

which is coincident with the Rayleigh criteria, when the total bandwidth is defined as the sum over the subband bandwidths.

The resolution limits derived from the ambiguity function do not depend on the *signal-to-noise ratio* (SNR). As in the Rayleigh criteria, the subliminal assumption of high enough SNR is assumed, as peaks in the output of the optimum filter can only be detected if they are over the noise floor. However, this method does not allow to quantify the effect of the SNR in the resolution performance.

The ambiguity function is also involved in accuracy bounds. Using again the Taylor series expansion for $|\chi(\tau, \nu)|^2$, the cut of the ambiguity function with a level plane close to the maximum can be expressed as an ellipse. A careful selection of the level plane leads to the so-called *uncertainty ellipse*, whose width along the τ-axis is

$$\delta\tau = \frac{1}{\beta\sqrt{2\mathcal{E}/\eta_0}} \tag{4.8}$$

where $\mathcal{E}$ is the signal energy, η_0 the noise power density and β the effective bandwidth defined in eq. (4.4). The parameter $\delta\tau$ gives the smallest possible estimation error in a root mean square sense.

4.3. Parameter estimation theory

The detection and ranging of objects in radar signals can be analyzed as a parameter estimation problem. Radar signals depend on the ranges and amplitudes of the present targets, so that special functions called estimators can be applied to the received signals in order to estimate the parameter of concern θ. The accuracy of the parameter estimate $\hat{\theta}$ depends on the noise present in the signal and on the quality of the estimator. They can be characterized by the mean and the variance of the estimate, which is a random variable. Estimators whose estimate coincides in mean with the parameter $E(\hat{\theta}) = \theta$ are called *unbiased* estimators, in opposite to the *biased* estimators, where the estimate mean presents a deviation from the parameter. Besides the mean value, the variance $\text{var}(\hat{\theta})$ is of interest, the lower the variance the more accurate is the estimate. The *Cramer-Rao lower bound* (CRLB) places a lower bound on the estimate variance, so that it permits to assert if an estimator is optimum or gives a benchmark against which the performance of any unbiased estimator can be compared. Furthermore, it alerts to the physical impossibility of finding an unbiased estimator whose variance is less than the bound.

The CRLB theorem for vector parameter estimation states that [37]:

$$\text{var}(\hat{\theta}_i) \geq \left[\mathbf{I}^{-1}(\boldsymbol{\theta})\right]_{i,i} \tag{4.9}$$

where $\boldsymbol{\theta}$ is the vector of M parameters to estimate

$$\boldsymbol{\theta} = \begin{pmatrix} \theta_1 & \theta_2 & \cdots & \theta_M \end{pmatrix}^T,$$

$\left[\mathbf{I}^{-1}(\boldsymbol{\theta})\right]_{i,i}$ are the diagonal elements of the inverse of the Fischer's information matrix $\mathbf{I}(\boldsymbol{\theta})$, whose element in the ith row and j^{th} column is

$$[\mathbf{I}(\boldsymbol{\theta})]_{i,j} = -E\left(\frac{\partial^2 \ln p(\mathbf{x};\boldsymbol{\theta})}{\partial\theta_i\partial\theta_j}\right) \tag{4.10}$$

and $p(\mathbf{x};\boldsymbol{\theta})$ is the *probability density function* (PDF) of the received signal $\mathbf{x}$ for a given set of parameters $\boldsymbol{\theta}$. This PDF considered as a function of the parameter set $\boldsymbol{\theta}$ is called the *likelihood function*.

The further discussion is restricted to complex data vectors $\mathbf{x}$ of length N

$$\mathbf{x} = \begin{pmatrix} x[0] & x[1] & \cdots & x[N-1] \end{pmatrix}^T$$

which have the complex Gaussian PDF

$$p(\mathbf{x};\boldsymbol{\theta}) = \frac{1}{\pi^N \det(\mathbf{C}_x(\boldsymbol{\theta}))} e^{-(\mathbf{x}-\mu(\boldsymbol{\theta}))^H \mathbf{C}_x^{-1}(\boldsymbol{\theta})(\mathbf{x}-\mu(\boldsymbol{\theta}))} \tag{4.11}$$

where the dependence on a vector of parameters is denoted by $\boldsymbol{\theta}$, $\mu(\boldsymbol{\theta})$ is the mean value $E(\mathbf{x}) = \mu(\boldsymbol{\theta})$, $\mathbf{C}_x(\boldsymbol{\theta})$ is the autocovariance matrix $E(\mathbf{x}^H\mathbf{x}) = \mathbf{C}_x(\boldsymbol{\theta})$, and all the parameters to estimate are real (complex parameters are separated into real and imaginary parts or into magnitude and phase). The expression for the Fischer information matrix elements results in

$$[\mathbf{I}(\boldsymbol{\theta})]_{i,j} = \text{tr}\left(\mathbf{C}_x^{-1}(\boldsymbol{\theta})\frac{\partial\mathbf{C}_x(\boldsymbol{\theta})}{\partial\theta_i}\mathbf{C}_x^{-1}(\boldsymbol{\theta})\frac{\partial\mathbf{C}_x(\boldsymbol{\theta})}{\partial\theta_j}\right)$$
$$+ 2\Re\left(\frac{\partial\mu^H(\boldsymbol{\theta})}{\partial\theta_i}\mathbf{C}_x^{-1}(\boldsymbol{\theta})\frac{\partial\mu(\boldsymbol{\theta})}{\partial\theta_j}\right) \tag{4.12}$$

where

$$\frac{\partial\mu(\boldsymbol{\theta})}{\partial\theta_i} = \begin{pmatrix} \frac{\partial[\mu(\boldsymbol{\theta})]_1}{\partial\theta_i} \\ \frac{\partial[\mu(\boldsymbol{\theta})]_2}{\partial\theta_i} \\ \vdots \\ \frac{\partial[\mu(\boldsymbol{\theta})]_N}{\partial\theta_i} \end{pmatrix}$$

$$\frac{\partial\mathbf{C}_x(\boldsymbol{\theta})}{\partial\theta_i} = \begin{pmatrix} \frac{\partial[\mathbf{C}_x(\boldsymbol{\theta})]_{1,1}}{\partial\theta_i} & \frac{\partial[\mathbf{C}_x(\boldsymbol{\theta})]_{1,2}}{\partial\theta_i} & \cdots & \frac{\partial[\mathbf{C}_x(\boldsymbol{\theta})]_{1,N}}{\partial\theta_i} \\ \frac{\partial[\mathbf{C}_x(\boldsymbol{\theta})]_{2,1}}{\partial\theta_i} & \frac{\partial[\mathbf{C}_x(\boldsymbol{\theta})]_{2,2}}{\partial\theta_i} & \cdots & \frac{\partial[\mathbf{C}_x(\boldsymbol{\theta})]_{2,N}}{\partial\theta_i} \\ \vdots & \vdots & \ddots & \vdots \\ \frac{\partial[\mathbf{C}_x(\boldsymbol{\theta})]_{N,1}}{\partial\theta_i} & \frac{\partial[\mathbf{C}_x(\boldsymbol{\theta})]_{N,2}}{\partial\theta_i} & \cdots & \frac{\partial[\mathbf{C}_x(\boldsymbol{\theta})]_{N,N}}{\partial\theta_i} \end{pmatrix}$$

and the derivative of a complex function is defined as:

$$\frac{\partial g}{\partial \theta_i} = \frac{\partial \Re(g)}{\partial \theta_i} + j \frac{\partial \Im(g)}{\partial \theta_i}.$$

Furthermore, the case of a deterministic data sequence $s[n; \boldsymbol{\theta}]$ received in the presence of additive *complex white Gaussian noise* (CWGN) with zero mean and variance σ_w^2, i.e. $w[n] \sim \mathcal{CN}(0, \sigma_w^2)$, is considered

$$x[n; \boldsymbol{\theta}] = s[n; \boldsymbol{\theta}] + w[n] \qquad n = 0, \ldots, N-1. \tag{4.13}$$

The mean value $\mu[n; \boldsymbol{\theta}]$ equals the signal $s[n; \boldsymbol{\theta}]$

$$\mu[n; \boldsymbol{\theta}] = s[n; \boldsymbol{\theta}] \tag{4.14}$$

and the autocovariance matrix $\mathbf{C}_x$ is independent of the parameter vector $\boldsymbol{\theta}$

$$\mathbf{C}_x = \sigma_w^2 \mathbf{I}_N, \tag{4.15}$$

so that the Fischer information matrix results in:

$$[\mathbf{I}(\boldsymbol{\theta})]_{i,j} = 2\Re \left(\frac{\partial \boldsymbol{\mu}^H(\boldsymbol{\theta})}{\partial \theta_i} \mathbf{C}_x^{-1} \frac{\partial \boldsymbol{\mu}(\boldsymbol{\theta})}{\partial \theta_j} \right) = \frac{2}{\sigma_w^2} \Re \left(\sum_{n=0}^{N-1} \frac{\partial s^*[n; \boldsymbol{\theta}]}{\partial \theta_i} \frac{\partial s[n; \boldsymbol{\theta}]}{\partial \theta_j} \right) \tag{4.16}$$

4.3.1. Application to radar accuracy

A first assessment of the best achievable accuracy in delay estimation can be done using the scalar version of the CRLB. Assume a signal $s(t+\tau)$ with finite energy $\mathcal{E}$ and received in the presence of additive white Gaussian noise with power density η_0. The only parameter to estimate is τ, so that the CRLB can be expressed as:

$$\mathrm{var}(\hat{\tau}) \geq \frac{1}{-E\left(\frac{\partial^2 \ln p(\mathbf{x}; \tau)}{\partial \tau^2} \right)} \tag{4.17}$$

and results in [37]:

$$\mathrm{var}(\hat{\tau}) = \delta \tau = \frac{1}{\beta \sqrt{2\mathcal{E}/\eta_0}} \tag{4.18}$$

where the effective bandwidth β is defined in eq. (4.4). Please note that the result is the same as in eq. (4.8), which was obtained with the ambiguity function.

Consider now the signal model described in chapter 2 with its poles laying on the unit circle

$$x[n] = s[n; \boldsymbol{\theta}] + w[n] = \sum_i h_i e^{j2\pi f_i n} + w[n] = \sum_i A_i e^{j\phi_i} e^{j2\pi f_i n} + w[n] \tag{4.19}$$

where each pole represents a scatterer at distance given by the normalized frequency f_i, the complex amplitude h_i is separated into magnitude A_i and phase ϕ_i (real parameters), and the parameter vector is composed by the amplitudes, phase terms and frequencies

$$\boldsymbol{\theta} = \begin{pmatrix} A_1 & \phi_1 & f_1 & \cdots & A_P & \phi_P & f_P \end{pmatrix}^T. \tag{4.20}$$

With the following expressions for the derivatives of the signal with respect to the parameters

$$\frac{\partial s[n; \boldsymbol{\theta}]}{\partial A_i} = e^{j\phi} e^{j2\pi f_i n} \tag{4.21a}$$

$$\frac{\partial s[n; \boldsymbol{\theta}]}{\partial \phi_i} = jA_i e^{j\phi} e^{j2\pi f_i n} \tag{4.21b}$$

$$\frac{\partial s[n; \boldsymbol{\theta}]}{\partial f_i} = j2\pi n A_i e^{j\phi} e^{j2\pi f_i n} \tag{4.21c}$$

$$\frac{\partial s^*[n; \boldsymbol{\theta}]}{\partial A_i} = e^{-j\phi} e^{-j2\pi f_i n} \tag{4.21d}$$

$$\frac{\partial s^*[n; \boldsymbol{\theta}]}{\partial \phi_i} = -jA_i e^{-j\phi} e^{-j2\pi f_i n} \tag{4.21e}$$

$$\frac{\partial s^*[n; \boldsymbol{\theta}]}{\partial f_i} = -j2\pi n A_i e^{-j\phi} e^{-j2\pi f_i n} \tag{4.21f}$$

the Fischer Information matrix and the CRLB for the different estimates can be obtained using equations (4.16) and (4.9).

Figure 4.1 shows the CRLB for frequency estimation in a two-tone signal:

$$s[n; \boldsymbol{\theta}] = A_1 e^{j\phi_1} e^{j2\pi f_1} + A_2 e^{j\phi_2} e^{j2\pi f_2}$$

where the parameter vector is

$$\boldsymbol{\theta} = \begin{pmatrix} A_1 & \phi_1 & f_1 & A_2 & \phi_2 & f_2 \end{pmatrix}^T.$$

It is observed that for increasing frequency distance Δf between the oscillations the estimation of their frequencies becomes more accurate independent of the absolute position of the two frequencies. Beyond 1.5 times the *Rayleigh* distance $(1/N)$ the variance of the estimates remains constant.

4.3.2. Conclusions

The behavior pattern of the CRLB for frequency estimation is coincident with the classic range resolution theory—i.e. with increasing Δf the resolution of the oscillations increases, and beyond the Rayleigh distance it remains constant—and therefore the CRLB has been proposed to estimate the resolution capability of radar signals, e.g. [43, 75, 85, 89, 97]. In this thesis

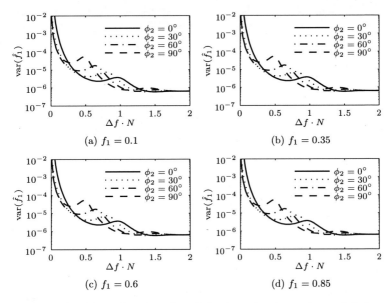

Figure 4.1: Cramer-Rao lower bound for the estimation variance of f_1 from a single-band signal with $N = 64$, SNR $= A_i^2/\sigma_w^2 = 0\,\mathrm{dB}$, $f_2 = f_1 + \Delta f$ and $\phi_1 = 0°$

the CRLB is only used to assess the range accuracy, the resolution problem is treated as a *detection* problem and not as an *estimation* problem. If the two oscillations in the radar signal are detected as two, the targets are said to be resolved, independent of how accurate they are ranged. Furthermore, there is a previous knowledge in the CRLB calculation, as the number of parameters, i. e. targets, is assumed to be known. The resolution problem—viewed as a discrimination problem between the cases of one or two targets present in the signal—is assumed to be solved when calculating the CRLB.

4.4. Hypothesis testing or detection theory

The resolution of a radar system can be defined as the capability to detect the actual presence of two targets instead of a single one (or none) in a noise corrupted signal. This problem can be formulated as an *hypothesis test* problem, i. e. the hypotheses of none, one or two targets present in the received signal are tested. This would be a *multiple hypothesis testing* as opposed to the *simple hypothesis testing*, where the detection of one signal is addressed. Hypothesis testing theory is extensively analyzed in the literature,

e. g. [38, 47, 48], and its application to radar is traditionally found in the design of optimum detectors—resulting in the matched filter detector for Gaussian noise [12, 55, 63, 83, 84, 96]—rather than to establish a limit on the resolution performance of a system, as is done here. Next, the principal theorems of hypothesis testing and their application to the radar resolution problem are presented.

4.4.1. Basic theory of hypotheses testing

Neyman-Pearson theorem

Consider two possible hypotheses: the *null hypothesis* $\mathcal{H}_0$ and the *alternative hypothesis* $\mathcal{H}_1$, and define the *probability of false alarm* as the probability to decide $\mathcal{H}_1$ when $\mathcal{H}_0$ is true[1]:

$$P_{\text{FA}} = P(\mathcal{H}_1; \mathcal{H}_0) \tag{4.22}$$

and the *probability of detection* as the probability to decide $\mathcal{H}_1$ when $\mathcal{H}_1$ is true

$$P_{\text{D}} = P(\mathcal{H}_1; \mathcal{H}_1). \tag{4.23}$$

To maximize P_{D} for a given $P_{\text{FA}} = \alpha$ decide $\mathcal{H}_1$ if

$$L(\mathbf{x}) = \frac{p(\mathbf{x}; \mathcal{H}_1)}{p(\mathbf{x}; \mathcal{H}_0)} > \gamma \tag{4.24}$$

where the threshold γ is found from

$$P_{\text{FA}} = \int\limits_{\mathbf{x}: L(\mathbf{x}) > \gamma} p(\mathbf{x}; \mathcal{H}_0) d\mathbf{x} = \alpha \tag{4.25}$$

and $p(\mathbf{x}; \mathcal{H}_i)$ represents the *probability density function* (PDF) of the random variable $\mathbf{x}$ when $\mathcal{H}_i$ is true. The function $L(\mathbf{x})$ is termed the *likelihood ratio test* (LRT).

Minimum probability of error

The *probability of error* is defined as:

$$P_{\text{e}} = P(\mathcal{H}_0 | \mathcal{H}_1) P(\mathcal{H}_1) + P(\mathcal{H}_1 | \mathcal{H}_0) P(\mathcal{H}_0), \tag{4.26}$$

where $P(\mathcal{H}_i | \mathcal{H}_j)$ is the *conditional* probability that indicates the probability of deciding $\mathcal{H}_i$ when $\mathcal{H}_j$ is true.

---

[1] Notation according to [38]

To minimize P_e decide $\mathcal{H}_1$ if

$$\frac{p(\mathbf{x}|\mathcal{H}_1)}{p(\mathbf{x}|\mathcal{H}_0)} > \frac{P(\mathcal{H}_0)}{P(\mathcal{H}_1)} = \gamma. \tag{4.27}$$

Again, the *conditional* likelihood ratio is compared to a threshold. For the equal probable case, $P(\mathcal{H}_0) = P(\mathcal{H}_1) = 1/2$, the hypothesis with maximum conditional likelihood is selected. This is called the *maximum likelihood* (ML) criteria

$$p(\mathbf{x}|\mathcal{H}_1) > p(\mathbf{x}|\mathcal{H}_0). \tag{4.28}$$

Using the Bayes rule

$$P(\mathcal{H}_i|\mathbf{x}) = \frac{p(\mathbf{x}|\mathcal{H}_i)P(\mathcal{H}_i)}{p(\mathbf{x})}, \tag{4.29}$$

eq. (4.27) can be rewritten as

$$P(\mathcal{H}_1|\mathbf{x}) > P(\mathcal{H}_0|\mathbf{x}), \tag{4.30}$$

which is termed the *maximum a posteriori* (MAP) criteria. The selected hypothesis has the maximum a posteriori (after the data is observed) probability.

Minimization of Bayes risk

A generalization of a minimum P_e criterion assigns cost to each type of error. The *Bayes risk* is defined as

$$\mathcal{R} = \sum_{i=0}^{1} \sum_{j=0}^{1} C_{ij} P(\mathcal{H}_i|\mathcal{H}_j) P(\mathcal{H}_j) \tag{4.31}$$

where C_{ij} is the cost if $\mathcal{H}_i$ is decided but $\mathcal{H}_j$ is true.

The detector that minimizes the Bayes risk $\mathcal{R}$ under the reasonable assumption that $C_{10} > C_{00}$, $C_{01} > C_{11}$, decides $\mathcal{H}_1$ if

$$\frac{p(\mathbf{x}|\mathcal{H}_1)}{p(\mathbf{x}|\mathcal{H}_0)} > \frac{(C_{10} - C_{00})P(\mathcal{H}_0)}{(C_{01} - C_{11})P(\mathcal{H}_1)} = \gamma. \tag{4.32}$$

Again, the conditional likelihood ratio is compared to a threshold. For the special case of $C_{00} = C_{11} = 0$ and $C_{01} = C_{10} = 1$ it can be seen that $\mathcal{R} = P_e$ and the criterion reduces to the ML.

Multiple hypothesis testing

Consider M hypothesis $\mathcal{H}_0, \mathcal{H}_1, \ldots, \mathcal{H}_{M-1}$. The Bayes risk is defined as

$$\mathcal{R} = \sum_{i=0}^{M-1} \sum_{j=0}^{M-1} C_{ij} P(\mathcal{H}_i|\mathcal{H}_j) P(\mathcal{H}_j), \tag{4.33}$$

which for the case

$$C_{ij} = \begin{cases} 0 & i = j \\ 1 & i \neq j \end{cases}$$

results in the probability of error $\mathcal{R} = P_e$.

The detector which minimizes $\mathcal{R}$ should choose the hypothesis which minimizes

$$C_i(\mathbf{x}) = \sum_{j=0}^{M-1} C_{ij} P(\mathcal{H}_j|\mathbf{x}) \tag{4.34}$$

For the case $\mathcal{R} = P_e$

$$C_i(\mathbf{x}) = \sum_{\substack{j=0 \\ j \neq i}}^{M-1} P(\mathcal{H}_j|\mathbf{x}) = \sum_{j=0}^{M-1} P(\mathcal{H}_j|\mathbf{x}) - P(\mathcal{H}_i|\mathbf{x})$$

and $C_i(\mathbf{x})$ minimizes when $P(\mathcal{H}_i|\mathbf{x})$ maximizes. The minimum P_e rule is to decide $\mathcal{H}_k$ if

$$P(\mathcal{H}_k|\mathbf{x}) > P(\mathcal{H}_i|\mathbf{x}) \qquad i \neq k, \tag{4.35}$$

which is the MAP criteria. For equal probable hypothesis

$$P(\mathcal{H}_i|\mathbf{x}) = \frac{p(\mathbf{x}|\mathcal{H}_i) P(\mathcal{H}_i)}{p(\mathbf{x})} = \frac{p(\mathbf{x}|\mathcal{H}_i) \frac{1}{M}}{p(\mathbf{x})}, \tag{4.36}$$

we decide $\mathcal{H}_k$ if

$$p(\mathbf{x}|\mathcal{H}_k) > p(\mathbf{x}|\mathcal{H}_i) \qquad i \neq k, \tag{4.37}$$

and the minimum P_e detector results in the ML decision rule.

4.4.2. Application to radar resolution

Assume a radar signal as described in chapter 2 sampled in the frequency domain at band $\mathcal{B}$

$$x[n] = s_i[n] + w[n] \qquad n = 0, \ldots, N-1 \tag{4.38}$$

where $w[n]$ is an additive *complex white Gaussian noise* (CWGN) with zero mean and variance σ_w^2, $w[n] \sim \mathcal{CN}(0, \sigma_w^2)$. The radar signal $s_i[n]$ corresponds to zero, one and two targets, which are also the hypothesis which should be tested, so that:

$$x[n] = s_0[n] + w[n] \qquad\qquad \mathcal{H}_0 \text{ no target} \tag{4.39a}$$
$$x[n] = s_1[n] + w[n] \qquad\qquad \mathcal{H}_1 \text{ one target} \tag{4.39b}$$
$$x[n] = s_2[n] + w[n] \qquad\qquad \mathcal{H}_2 \text{ two targets} \tag{4.39c}$$

with

$$s_0[n] = 0$$
$$s_1[n] = h_m e^{j2\pi f_m n} = A_m e^{j\phi_m} e^{j2\pi f_m n}$$
$$s_2[n] = h_1 e^{j2\pi f_1 n} + h_2 e^{j2\pi f_2 n} = A_1 e^{j\phi_1} e^{j2\pi f_1 n} + A_2 e^{j\phi_2} e^{j2\pi f_2 n}.$$

The *probability of resolution* is defined as the probability to detect two targets when two targets are present, i.e. to select hypothesis $\mathcal{H}_2$ when $\mathcal{H}_2$ is true:

$$P_{\mathrm{R}} = P(\mathcal{H}_2 | \mathcal{H}_2). \tag{4.40}$$

The P_{R} of two different optimum detectors based on different signal knowledge is presented next. First, the optimum detector when the received signal is a perfectly known deterministic data sequence is presented, followed by the detector for random radar signals which uses perfect knowledge of their *probability density function* (PDF). Neither of these two assumptions is fulfilled in a real radar case, as nor the received signal neither several parameters of its PDF, e.g. f_i, are previously known. A detector that assumes perfect knowledge of an unknown parameter to design an optimum detector is referred as a *clairvoyant detector* and its performance is used as an upper bound, an analogous situation to the use of the Cramer-Rao lower bound for the variance of an estimation.

Optimum detector for known deterministic radar signals

The received radar signals are perfectly known deterministic data sequences with additive CWGN. The ML criteria is applied to minimize P_{e}. The likelihood function is

$$L_{\mathbf{x}}(\mathcal{H}_i) = p(\mathbf{x}|\mathcal{H}_i) = \frac{1}{\pi^N \sigma_w^{2N}} e^{-\frac{1}{\sigma_w^2}(\mathbf{x}-\mathbf{s}_i)^H(\mathbf{x}-\mathbf{s}_i)} \tag{4.41}$$

where

$$\mathbf{x} = \begin{pmatrix} x[0] & \dots & x[N-1] \end{pmatrix}^T$$
$$\mathbf{s}_i = \begin{pmatrix} s_i[0] & \dots & s_i[N-1] \end{pmatrix}^T.$$

Taking the logarithm (a monotonically increasing transformation)

$$\ln(L_\mathbf{x}(\mathcal{H}_i)) = l_\mathbf{x}(\mathcal{H}_i) = \ln\left(\frac{1}{\pi^N \sigma_w^{2N}}\right) - \frac{1}{\sigma_w^2}(\mathbf{x} - \mathbf{s}_i)^H(\mathbf{x} - \mathbf{s}_i)$$

and suppressing common terms, the selected hypothesis will maximize the test

$$T_i(\mathbf{x}) = \Re(\mathbf{s}_i^H \mathbf{x}) - \frac{1}{2}\mathbf{s}_i^H \mathbf{s}_i = \Re(\mathbf{s}_i^H \mathbf{x}) - \frac{1}{2}\mathcal{E}_i \qquad (4.42)$$

which is a Gaussian variable.

The P_R is the $P(\mathcal{H}_2|\mathcal{H}_2) = P(T_2 > T_1, T_2 > T_0|\mathcal{H}_2)$. Taking into account that $T_0 = 0$, $\mathcal{H}_2$ is selected when $T_2 > \max(T_1, 0)$. This is an order statistics problem. For non-orthogonal signals this is an analytically intractable problem. To obtain analytical results, only the discrimination between the two hypotheses $\mathcal{H}_1$ and $\mathcal{H}_2$—the hypothesis where no target is present is rejected— is considered. For this case $\mathcal{H}_2$ is selected if $T_2(\mathbf{x}) > T_1(\mathbf{x})$ and the P_R can be expressed as:

$$\begin{aligned} P_\mathrm{R} &= P(T_2(\mathbf{x}) > T_1(\mathbf{x})|\mathcal{H}_2) \\ &= P(T_2(\mathbf{x}) - T_1(\mathbf{x}) > 0|\mathcal{H}_2) = P(T(\mathbf{x}) > 0|\mathcal{H}_2) \end{aligned} \qquad (4.43)$$

with

$$T(\mathbf{x}) = T_2(\mathbf{x}) - T_1(\mathbf{x}) = \Re((\mathbf{s}_2 - \mathbf{s}_1)^H \mathbf{x}) - \frac{1}{2}(\mathcal{E}_2 - \mathcal{E}_1). \qquad (4.44)$$

$T(\mathbf{x})$ is a Gaussian variable with moments

$$E(T(\mathbf{x})|\mathcal{H}_2) = \frac{1}{2}(\mathbf{s}_2 - \mathbf{s}_1)^H(\mathbf{s}_2 - \mathbf{s}_1)$$

$$\mathrm{var}(T(\mathbf{x})|\mathcal{H}_2) = \frac{\sigma_w^2}{2}(\mathbf{s}_2 - \mathbf{s}_1)^H(\mathbf{s}_2 - \mathbf{s}_1)$$

where it is used that the real and imaginary parts of a complex Gaussian variable are real Gaussian variables, independent of each other and with the same variance (one-half the variance of the complex Gaussian variable).

The probability of resolution can be written as

$$P_\mathrm{R} = P(T(\mathbf{x}) > 0) = Q\left(-\sqrt{\frac{(\mathbf{s}_2 - \mathbf{s}_1)^H(\mathbf{s}_2 - \mathbf{s}_1)}{2\sigma_w^2}}\right) \qquad (4.45)$$

where Q is the complementary cumulative distribution function of a Gaussian variable [1, 9]. Defining $\mathcal{E}_d$ as the energy of the difference signal $(\mathbf{s}_2 - \mathbf{s}_1)$ when

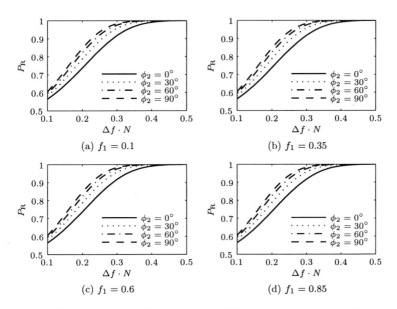

Figure 4.2: Resolution performance of the ML detector for deterministic radar signals. For all cases $N = 64$, SNR = 0 dB, $f_2 = f_1 + \Delta f$ and $\phi_1 = 0°$

$A_1 = A_2 = A = 1$ and the signal-to-noise ratio as SNR $= A^2/\sigma_w^2$, P_R can be rewritten as:

$$P_R = Q\left(-\sqrt{\text{SNR}\frac{\mathcal{E}_d}{2}}\right). \tag{4.46}$$

Analyzing this expression, it can be stated that two signals are more likely to be resolved by increasing the SNR or the energy of the difference signal. The increase of $\mathcal{E}_d$ to predict a better resolution performance has also been used in the literature [63] to justify the use of the ambiguity function.

To assess the P_R the following two-target signal is considered

$$s_2[n] = e^{j\phi_1}e^{j2\pi f_1 n} + e^{j\phi_2}e^{j2\pi f_2 n}$$

and as one-target signal, the sequence

$$s_1[n] = A_m e^{j\phi_m}e^{j2\pi f_m n}$$

with the parameters f_m, ϕ_m and A_m which minimize $\mathcal{E}_d$ is selected.

Figure 4.2 shows the P_R as function of the frequency difference $\Delta f = f_2 - f_1$. As expected, the resolution performance improves as Δf increases, independent of the value of the center frequency $(f_1 + f_2)/2$. This result is consistent with those obtained with classical approaches. In this analysis, only the one- and two-targets hypotheses have been considered. The inclusion of the $\mathcal{H}_0$ would mean an increase in the necessary SNR to discard the null hypothesis. The resolution behavior, however, would be similar.

Optimum detector for random radar signals with known PDF

The received radar signals are random data sequences with known PDF. For mathematical simplicity, the complex amplitudes of the received exponentials are considered independent random complex Gaussian variables with zero mean $h_1 \sim \mathcal{CN}(0, \sigma^2)$, $h_2 \sim \mathcal{CN}(0, \sigma^2)$ and $h_m \sim \mathcal{CN}(0, \sigma_m^2)$, so that their phases are uniformly distributed, $\phi_i \sim \mathcal{U}(-\pi, \pi)$, and their absolute values have a Rayleigh distribution with mean and variance

$$E(A_i) = \sqrt{\frac{\pi \sigma_i^2}{4}} \tag{4.47a}$$

$$\text{var}(A_i) = \left(1 - \frac{\pi}{4}\right) \sigma_i^2. \tag{4.47b}$$

The received radar signals have also a complex Gaussian distribution with zero mean $\mathbf{x} \sim \mathcal{CN}(0, \mathbf{C}_{s_i} + \sigma_w^2 \mathbf{I})$, where $\mathbf{C}_{s_i}$ is the *autocorrelation matrix* of the signal $s_i[n]$—corresponding to hypothesis $\mathcal{H}_i$—defined as

$$\mathbf{C}_{s_i} = E(\mathbf{s}_i \cdot \mathbf{s}_i^H) = \begin{pmatrix} r_{s_i s_i}[0, 0] & \cdots & r_{s_i s_i}[0, -N+1] \\ \vdots & \ddots & \vdots \\ r_{s_i s_i}[N-1, N-1] & \cdots & r_{s_i s_i}[N-1, 0] \end{pmatrix} \tag{4.48}$$

with $\mathbf{s}_i$ defined as in eq. (4.41), and $r_{s_i s_i}$ being the *autocorrelation sequence* (ACS)

$$r_{s_1 s_1}[n, n+m] = \sigma_m^2 e^{j2\pi f_m m} \tag{4.49a}$$

$$r_{s_2 s_2}[n, n+m] = \sigma^2 e^{j2\pi f_1 m} + \sigma^2 e^{j2\pi f_2 m}. \tag{4.49b}$$

The ML criteria is applied to minimize the *probability of error* P_e. As in the case where the received signals were known, the emphasis is placed on the discrimination case, so that hypothesis $\mathcal{H}_1$ and $\mathcal{H}_2$ are to be tested. The likelihood function is

$$L_\mathbf{x}(\mathcal{H}_i) = p(\mathbf{x}|\mathcal{H}_i) = \frac{1}{\pi^N \det\left(\mathbf{C}_{s_i} + \sigma_w^2 \mathbf{I}\right)} e^{-\mathbf{x}^H \left(\mathbf{C}_{s_i} + \sigma_w^2 \mathbf{I}\right)^{-1} \mathbf{x}}.$$

Taking the logarithm

$$\ln(L_\mathbf{x}(\mathcal{H}_i)) = l_\mathbf{x}(\mathcal{H}_i) = \ln\left(\frac{1}{\pi^N \det(\mathbf{C}_{s_i} + \sigma_w^2 \mathbf{I})}\right) - \mathbf{x}^H \left(\mathbf{C}_{s_i} + \sigma_w^2 \mathbf{I}\right)^{-1} \mathbf{x}$$

and suppressing common terms, the selected hypothesis will maximize the test function

$$T_i(\mathbf{x}) = -\ln\left(\det\left(\mathbf{C}_{s_i} + \sigma_w^2\mathbf{I}\right)\right) - \mathbf{x}^H\left(\mathbf{C}_{s_i} + \sigma_w^2\mathbf{I}\right)^{-1}\mathbf{x} \qquad (4.50)$$

In the test functions $T_i(\mathbf{x})$ two $N \times N$ matrices have to be inverted. To reduce the computational effort, an alternative test function can be developed based on the decomposition of the vector $\mathbf{s}_i$ as the product of a constant matrix $\mathbf{H}_i$ and a random vector $\boldsymbol{\theta}_i$

$$\mathbf{s}_1 = \mathbf{H}_1 \cdot \boldsymbol{\theta}_1 = \begin{pmatrix} 1 \\ e^{j\omega_m} \\ \vdots \\ e^{j\omega_m(N-1)} \end{pmatrix} \cdot \left(h_m\right) \qquad (4.51a)$$

$$\mathbf{s}_2 = \mathbf{H}_2 \cdot \boldsymbol{\theta}_2 = \begin{pmatrix} 1 & 1 \\ e^{j\omega_1} & e^{j\omega_2} \\ \vdots & \vdots \\ e^{j\omega_1(N-1)} & e^{j\omega_2(N-1)} \end{pmatrix} \cdot \begin{pmatrix} h_1 \\ h_2 \end{pmatrix} \qquad (4.51b)$$

so that

$$\mathbf{C}_{s_i} = \mathbf{H}_i\mathbf{C}_{\theta_i}\mathbf{H}_i^H$$

where $\mathbf{C}_{\theta_i} = E(\boldsymbol{\theta}_i\boldsymbol{\theta}_i^H)$. Taking into account that

$$\mathbf{C}_{\theta_1} = \sigma_m^2 \qquad (4.52a)$$
$$\mathbf{C}_{\theta_2} = \sigma^2\mathbf{I}_2 \qquad (4.52b)$$

the test function can be reduced to

$$T_1(\mathbf{x}) = -\ln\left(\frac{\sigma_m^2}{\sigma_w^2}\mathbf{H}_1^H\mathbf{H}_1 + 1\right)$$
$$\qquad\qquad - \frac{1}{\sigma_w^2}\mathbf{x}^H\left(\mathbf{I}_N - \frac{\mathbf{H}_1\mathbf{H}_1^H}{\frac{\sigma_w^2}{\sigma_m^2} + \mathbf{H}_1^H\mathbf{H}_1}\right)\mathbf{x} \qquad (4.53a)$$

$$T_2(\mathbf{x}) = -\ln\left(\det\left(\frac{\sigma^2}{\sigma_w^2}\mathbf{H}_2^H\mathbf{H}_2 + \mathbf{I}_2\right)\right)$$
$$\qquad\qquad - \frac{1}{\sigma_w^2}\mathbf{x}^H\left(\mathbf{I}_N - \mathbf{H}_2\left(\frac{\sigma_w^2}{\sigma^2}\mathbf{I}_2 + \mathbf{H}_2^H\mathbf{H}_2\right)^{-1}\mathbf{H}_2^H\right)\mathbf{x} \qquad (4.53b)$$

where common terms have been suppressed, and the determinant property

$$\det\left(\mathbf{A}\mathbf{A}^H + \mathbf{I}\right) = \det\left(\mathbf{A}^H\mathbf{A} + \mathbf{I}\right)$$

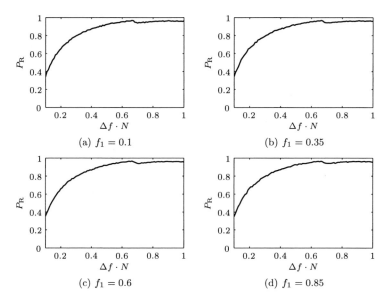

Figure 4.3: Resolution performance of the ML detector for random signals with known Gaussian PDF. For all cases $N = 64$, SNR $= 10\,\mathrm{dB}$, $f_2 = f_1 + \Delta f$

and the matrix inversion lemma

$$(\mathbf{A} + \mathbf{UCV})^{-1} = \mathbf{A}^{-1} - \mathbf{A}^{-1}\mathbf{U}(\mathbf{C}^{-1} + \mathbf{VA}^{-1}\mathbf{U})^{-1}\mathbf{VA}^{-1}$$

have been used. Now only a 2×2 matrix has to be inverted.

As there is no analytical expression for P_R a Monte Carlo analysis is necessary. The following ACS for the $\mathcal{H}_2$ hypothesis is considered

$$r_{s_2 s_2}[m] = \sigma^2 e^{j2\pi f_1 m} + \sigma^2 e^{j2\pi f_2 m}$$

which corresponds to a signal with two tones with independent complex Gaussian amplitudes with equal variance σ^2. The signal-to-noise ratio is defined as SNR $= \sigma^2/\sigma_w^2$. For the one-target signal, the following one-tone ACS

$$r_{s_1 s_1}[m] = \sigma_m^2 e^{j2\pi f_m m}$$

is optimized over f_m and σ_m^2 to minimize the difference—in a quadratic error sense—to $r_{s_2 s_2}[m]$. The one-target ACS which is closer to the actual two-target ACS is selected to obtain a bound for P_R. The random two-target signal is evaluated with the test corresponding to the sequence which—in mean sense—minimizes the difference between the hypothesis sequences.

Figure 4.3 shows the P_R as function of the frequency difference $\Delta f = f_2 - f_1$ obtained with 10^4 trials for each point. As test signal, two complex oscillations with independent random complex Gaussian amplitudes are used. As in the case for deterministic radar signals the resolution performance improves as Δf increases independent of the value of the center frequency $(f_1 + f_2)/2$. As expected, the resolution performance of this detector is poorer than for the previous case, as less information of the signal is available. Although $\mathcal{H}_0$ is not included in the analysis, the resolution behavior would remain similar after its inclusion. It would only mean an increase in the necessary SNR to discard the null hypothesis.

4.4.3. Conclusions

The hypothesis testing theory is a good approach to assert the resolution limits in a radar system. The obtained results have been shown to be consistent with those of classical approaches, i.e. resolution becomes more probable with increasing target distance—frequency distance in the spectral analysis case—and beyond the Rayleigh bound remains constant. The used test functions are the so-called clairvoyant detectors. They assume an unavailable knowledge of the signal or its PDF and show an upper bound in the resolution performance. The resolution in specific radar systems typically falls below the clairvoyant detector resolution limit, although their qualitative behaviors would be similar.

5. Range accuracy in multiband radar systems

This chapter deals with the range estimation performance which can be achieved in dualband radar systems, a particular case of multiband systems. The obtained results, however, can be extended to the more general case. The *Cramer-Rao lower bound* (CRLB) for parameter estimation presented in chapter 4 is applied to *coherent* and *non-coherent* dualband radar systems.

5.1. Parameter estimation theory in dualband radar systems

To assess the possibility of increasing the range accuracy in dualband radar systems, the CRLB is applied here to radar signals evaluated at two different observation windows, i. e. frequency bands. Coherent and non-coherent systems are analyzed, where the relative distance between the observation windows is known or unknown, respectively.

Consider a deterministic signal sampled at windows $\mathcal{B}_1$ and $\mathcal{B}_2$ in the presence of additive *complex white Gaussian noise* (CWGN) $w[n] \sim \mathcal{CN}(0, \sigma_w^2)$

$$x[n; \boldsymbol{\theta}] = s[n; \boldsymbol{\theta}] + w[n] \qquad n = \begin{cases} N_{10}, \ldots, N_{10} + N_1 - 1 & \text{at } \mathcal{B}_1 \\ N_{20}, \ldots, N_{20} + N_2 - 1 & \text{at } \mathcal{B}_2 \end{cases} \qquad (5.1)$$

where the dependence on a parameter set is denoted by $\boldsymbol{\theta}$. The deterministic signal is, in this case, a radar signal in the frequency domain as described in chapter 2, i. e. a sum of P complex exponentials where each pole represents a scatterer at a distance given by the normalized frequency f_i

$$s[n; \boldsymbol{\theta}] = \sum_{i=1}^{P} h_i e^{j2\pi f_i n} = \sum_{i=1}^{P} A_i e^{j\phi_i} e^{j2\pi f_i n}. \qquad (5.2)$$

The parameter vector is composed by the real amplitudes, phase terms and frequencies of the exponentials

$$\boldsymbol{\theta} = \begin{pmatrix} A_1 & \phi_1 & f_1 & \cdots & A_P & \phi_P & f_P \end{pmatrix}^T \qquad (5.3)$$

where all the elements are real, as the complex amplitude is separated into magnitude and phase. In chapter 4 the expression of the CRLB for vector estimates of deterministic signals with additive CWGN was presented:

$$\text{var}(\hat{\theta}_i) \geq \left[\mathbf{I}^{-1}(\boldsymbol{\theta}) \right]_{i,i} \qquad (5.4)$$

where the Fischer's information matrix for deterministic signals has the expression

$$[\mathbf{I}(\boldsymbol{\theta})]_{i,j} = \frac{2}{\sigma_w^2} \Re \left(\sum_n \frac{\partial s^*[n; \boldsymbol{\theta}]}{\partial \theta_i} \frac{\partial s[n; \boldsymbol{\theta}]}{\partial \theta_j} \right) \tag{5.5}$$

It is observed that the Fischer information matrix is constructed summing over the independent variable n, where the multiband information is included. In *coherent* systems, the gap $\Delta N = N_{20} - N_{10} + N_1$ between the observation windows $\mathcal{B}_1$ and $\mathcal{B}_2$ is exactly known, so that the expression for the index n in eq. (5.1) can be used directly. In *non-coherent* systems, however, ΔN is unknown, so that it is considered as another parameter on which the received signal depends, and some modifications to obtain the Fischer information matrix are required. Next, results for the coherent and the non-coherent cases are presented.

5.1.1. Coherent system

The expression in eq. (5.5) for the Fischer informations matrix and the derivatives in eq. (4.21) are used to evaluate the CRLB in a two tone signal defined as:

$$s[n; \boldsymbol{\theta}] = A_1 e^{j\phi_1} e^{j2\pi f_1 n} + A_2 e^{j\phi_2} e^{j2\pi f_2 n}$$

which depends on the vector parameter

$$\boldsymbol{\theta} = \begin{pmatrix} A_1 & \phi_1 & f_1 & A_2 & \phi_2 & f_2 \end{pmatrix}^T$$

and is received in the presence of additive CWGN $w[n] \sim \mathcal{CN}(0, \sigma_w^2)$. The dualband information is included in the index n defined as in eq. (5.1)

$$n = \begin{cases} N_{10}, \ldots, N_{10} + N_1 - 1 & \text{at } \mathcal{B}_1 \\ N_{20}, \ldots, N_{20} + N_2 - 1 & \text{at } \mathcal{B}_2 \end{cases}$$

Figure 5.1 shows the CRLB for the frequency estimation variance as a function of the gap ΔN between the observation windows $\mathcal{B}_1$ and $\mathcal{B}_2$. As in the single-band case, the CRLB depends on the frequency difference $\Delta f = f_2 - f_1$ but not on the center frequency $(f_1 + f_2)/2$, so that only one f_1 case for each ΔN has been shown. It is observed that for all cases the estimation variance decreases with increasing window gap ΔN, i.e. a coherent dualband radar system with constant subband bandwidths becomes more accurate when the band gap between them increases. The estimation variance presents for $\Delta f = 2/(N_1 + N_2)$ a constant decrease with the window gap. For lower frequency distances $\Delta f < 1/(N_1 + N_2)$, however, local minima appear at periodic

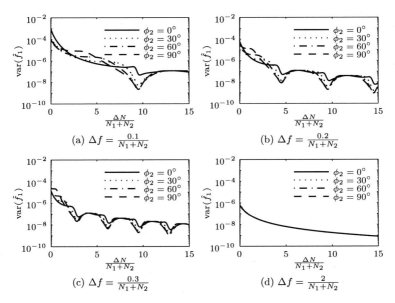

Figure 5.1: Cramer-Rao lower bound for the estimation variance of f_1 from a dualband coherent signal with $N_1 = N_2 = 32$, SNR $= A_i/\sigma_w^2 = 0\,\text{dB}$, $f_1 = 0.1$, $\phi_1 = 0°$, $f_2 = f_1 + \Delta f$ and variable gap ΔN

distances $1/\Delta f$. Rewriting the sum of two exponentials as

$$s[n] = e^{j(2\pi f_1 n + \phi_1)} + e^{j(2\pi f_2 n + \phi_2)} =$$
$$= 2e^{j\left(2\pi \frac{f_1 + f_2}{2} n + \frac{\phi_1 + \phi_2}{2}\right)} \cos\left(2\pi \frac{f_1 - f_2}{2} n + \frac{\phi_1 - \phi_2}{2}\right) \quad (5.6)$$

it is observed that the periodicity of the minima in the estimate variance is half a period of the cosinus term—the envelope of the complex exponential—in the previous expression. This periodic effect is related to the correlation between the two exponentials. For Δf below the Rayleigh bound, depending on the actual phase at the evaluation windows, the correlation between the two tones varies and influences the parameter estimate. For distances over the Rayleigh bound, however, the two exponentials can be considered orthogonal, i. e. no influence of the actual phase or observation windows is expected. This is observed in figure 5.1d, where the actual phase terms have no influence on the estimate variance.

5.1.2. Non-coherent system

In a dualband non-coherent system, the window gap is not exactly known, i. e. there is an additional unknown parameter to estimate in the sequence index n. In order to separate the unknown parameter from the sequence index, it can be rewritten as

$$
n = \begin{cases} N_c - \frac{\Delta N'}{2} + k_1 & \text{at } \mathcal{B}_1 \\ N_c + \frac{\Delta N'}{2} + k_2 & \text{at } \mathcal{B}_2 \end{cases} \tag{5.7}
$$

where

$$
\begin{aligned}
\Delta N' &= N_{20} - N_{10} \\
k_1 &= 0, \ldots, N_1 - 1 \\
k_2 &= 0, \ldots, N_2 - 1
\end{aligned}
$$

and $\Delta N'$ is the additional parameter to estimate. The signal $s[n; \boldsymbol{\theta}]$ is decomposed into the subwindow signals

$$
s_1[k_1; \boldsymbol{\theta}] = \sum_{i=1}^{P} A_i e^{j\phi_i} e^{j2\pi f_i (N_c - \frac{\Delta N'}{2} + k_1)} \tag{5.8a}
$$

$$
s_2[k_2; \boldsymbol{\theta}] = \sum_{i=1}^{P} A_i e^{j\phi_i} e^{j2\pi f_i (N_c + \frac{\Delta N'}{2} + k_2)} \tag{5.8b}
$$

with k_l as sequence index and parameter vector

$$
\boldsymbol{\theta} = \begin{pmatrix} A_1 & \phi_1 & f_1 & \cdots & A_P & \phi_P & f_P & \Delta N' \end{pmatrix}^T \tag{5.9}
$$

The Fischer information matrix is then rewritten as:

$$
[\mathbf{I}(\boldsymbol{\theta})]_{i,j} = \frac{2}{\sigma_w^2} \Re \left(\sum_{k_1} \frac{\partial s_1^*[k_1; \boldsymbol{\theta}]}{\partial \theta_i} \frac{\partial s_1[k_1; \boldsymbol{\theta}]}{\partial \theta_j} + \sum_{k_2} \frac{\partial s_2^*[k_2; \boldsymbol{\theta}]}{\partial \theta_i} \frac{\partial s_2[k_2; \boldsymbol{\theta}]}{\partial \theta_j} \right) \tag{5.10}
$$

The derivatives of the signal with respect to the different parameters are now required, a list follows.

$$
\frac{\partial s_1[k_1; \boldsymbol{\theta}]}{\partial A_i} = e^{j\phi_i} e^{j2\pi f_i (N_c - \frac{\Delta N'}{2} + k_1)} \tag{5.11a}
$$

$$
\frac{\partial s_1[k_1; \boldsymbol{\theta}]}{\partial \phi_i} = j A_i e^{j\phi_i} e^{j2\pi f_i (N_c - \frac{\Delta N'}{2} + k_1)} \tag{5.11b}
$$

$$
\frac{\partial s_1[k_1; \boldsymbol{\theta}]}{\partial f_i} = j2\pi (N_c - \frac{\Delta N'}{2} + k_1) A_i e^{j\phi_i} e^{j2\pi f_i (N_c - \frac{\Delta N'}{2} + k_1)} \tag{5.11c}
$$

$$
\frac{\partial s_1[k_1; \boldsymbol{\theta}]}{\partial \Delta N'} = \sum_i -j2\pi f_i \frac{1}{2} A_i e^{j\phi_i} e^{j2\pi f_i (N_c - \frac{\Delta N'}{2} + k_1)} \tag{5.11d}
$$

$$\frac{\partial s_1^*[k_1; \boldsymbol{\theta}]}{\partial A_i} = e^{-j\phi_i} e^{-j2\pi f_i (N_c - \frac{\Delta N'}{2} + k_1)} \tag{5.11e}$$

$$\frac{\partial s_1^*[k_1; \boldsymbol{\theta}]}{\partial \phi_i} = -j A_i e^{-j\phi_i} e^{-j2\pi f_i (N_c - \frac{\Delta N'}{2} + k_1)} \tag{5.11f}$$

$$\frac{\partial s_1^*[k_1; \boldsymbol{\theta}]}{\partial f_i} = -j2\pi (N_c - \frac{\Delta N'}{2} + k_1) A_i e^{-j\phi_i} e^{-j2\pi f_i (N_c - \frac{\Delta N'}{2} + k_1)} \tag{5.11g}$$

$$\frac{\partial s_1^*[k_1; \boldsymbol{\theta}]}{\partial \Delta N'} = \sum_i j2\pi f_i \frac{1}{2} A_i e^{-j\phi_i} e^{-j2\pi f_i (N_c - \frac{\Delta N'}{2} + k_1)} \tag{5.11h}$$

$$\frac{\partial s_2[k_2; \boldsymbol{\theta}]}{\partial A_i} = e^{j\phi_i} e^{j2\pi f_i (N_c + \frac{\Delta N'}{2} + k_2)} \tag{5.11i}$$

$$\frac{\partial s_2[k_2; \boldsymbol{\theta}]}{\partial \phi_i} = j A_i e^{j\phi_i} e^{j2\pi f_i (N_c + \frac{\Delta N'}{2} + k_2)} \tag{5.11j}$$

$$\frac{\partial s_2[k_2; \boldsymbol{\theta}]}{\partial f_i} = j2\pi (N_c + \frac{\Delta N'}{2} + k_2) A_i e^{j\phi_i} e^{j2\pi f_i (N_c + \frac{\Delta N'}{2} + k_2)} \tag{5.11k}$$

$$\frac{\partial s_2[k_2; \boldsymbol{\theta}]}{\partial \Delta N'} = \sum_i j2\pi f_i \frac{1}{2} A_i e^{j\phi_i} e^{j2\pi f_i (N_c + \frac{\Delta N'}{2} + k_2)} \tag{5.11l}$$

$$\frac{\partial s_2^*[k_2; \boldsymbol{\theta}]}{\partial A_i} = e^{-j\phi_i} e^{-j2\pi f_i (N_c + \frac{\Delta N'}{2} + k_2)} \tag{5.11m}$$

$$\frac{\partial s_2^*[k_2; \boldsymbol{\theta}]}{\partial \phi_i} = -j A_i e^{-j\phi_i} e^{-j2\pi f_i (N_c + \frac{\Delta N'}{2} + k_2)} \tag{5.11n}$$

$$\frac{\partial s_2^*[k_2; \boldsymbol{\theta}]}{\partial f_i} = -j2\pi (N_c + \frac{\Delta N'}{2} + k_2) A_i e^{-j\phi_i} e^{-j2\pi f_i (N_c + \frac{\Delta N'}{2} + k_2)} \tag{5.11o}$$

$$\frac{\partial s_2^*[k_2; \boldsymbol{\theta}]}{\partial \Delta N'} = \sum_i -j2\pi f_i \frac{1}{2} A_i e^{-j\phi_i} e^{-j2\pi f_i (N_c + \frac{\Delta N'}{2} + k_2)} \tag{5.11p}$$

This expressions are now applied to evaluate the CRLB in a two tone signal observed by a non-coherent dualband radar system:

$$s_l[k_l; \boldsymbol{\theta}] = A_1 e^{j\phi_1} e^{j2\pi f_1 \left(N_c \pm \frac{\Delta N'}{2} + k_l\right)} + A_2 e^{j\phi_2} e^{j2\pi f_2 \left(N_c \pm \frac{\Delta N'}{2} + k_l\right)}$$

where

$$\boldsymbol{\theta} = \begin{pmatrix} A_1 & \phi_1 & f_1 & A_2 & \phi_2 & f_2 & \Delta N' \end{pmatrix}^T$$

and is received in the presence of additive CWGN, $w[n] \sim \mathcal{CN}(0, \sigma_w^2)$.

Figure 5.2 shows the CRLB for the frequency estimation variance as a function of the gap ΔN between the observation windows $\mathcal{B}_1$ and $\mathcal{B}_2$. As in the single-band and the coherent cases, the CRLB depends on the frequency difference $\Delta f = f_2 - f_1$ but not on the center frequency $(f_1 + f_2)/2$. In opposite to the coherent case, the estimation variance does not decrease with increasing

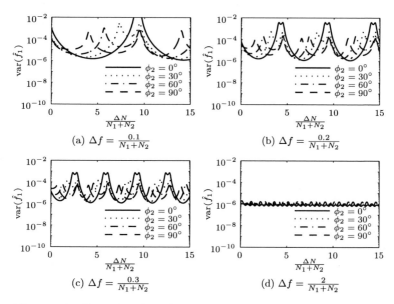

Figure 5.2: Cramer-Rao lower bound for the estimation variance of f_1 from a dualband non-coherent signal with $N_1 = N_2 = 32$, $\text{SNR} = A_i^2/\sigma_w^2 = 0\,\text{dB}$, $f_1 = 0.1$, $\phi_1 = 0°$, $f_2 = f_1 + \Delta f$ and variable gap ΔN

window gap ΔN, i.e. a non-coherent dualband radar system does not improve, in this case, its range accuracy by increasing the gap between the subbands. A periodic behavior is observed in the CRLB for the frequency estimate for $\Delta f < 1/(N_1 + N_2)$. This effect was also observed and discussed in the coherent case: the lack of orthogonality between the two tones for frequency distances below the Rayleigh bound makes the estimation dependent on the actual phase terms of the oscillations at the observation windows. The period coincides with a half period of the cosinus term in the two-tone decomposition shown in eq. (5.6).

Comparing the results of the coherent and the non-coherent systems, it is clear that the lack of knowledge of the gap ΔN is responsible for the poorer performance of the range accuracy in the non-coherent case. It is worth to test if including more exponentials in the signal $s[n; \boldsymbol{\theta}]$ compensates this effect. As each exponential in the signal is affected in the same way by ΔN, the more exponentials are present in the signal, the more accurate ΔN can be estimated. Figures 5.3 and 5.4 show the estimate variance of the frequency f_1 and the

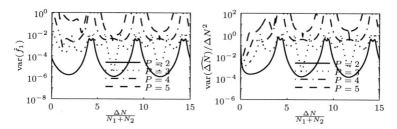

Figure 5.3: Cramer-Rao lower bound for the estimation variance of f_1 and ΔN from a dualband non-coherent signal with $N_1 = N_2 = 32$, SNR $= A_i^2/\sigma_w^2 = 0\,\mathrm{dB}$, variable gap ΔN and P exponentials with $A_i = 1$, $\phi_i = 0°$, $f_1 = 0.1$ and $f_i = f_{i-1} + \frac{0.2}{N_1+N_2}$

estimate variance of ΔN for signals with the following expression:

$$s_l[k_l; \boldsymbol{\theta}] = \sum_{i=1}^{P} A_i e^{j\phi_i} e^{j2\pi(f_1+(i-1)\Delta f)(N_c \pm \frac{\Delta N'}{2}+k_l)}.$$

It can be observed that the accuracy of the frequency and window gap estimates increase with the number of exponentials P only if the frequency distance between them is large enough, i.e. over the Rayleigh bound. For lower distances a decrease in the accuracy for increasing P is observed. The next step is to obtain the CRLB for a signal with two closely spaced sinusoids

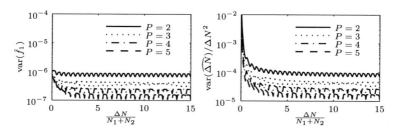

Figure 5.4: Cramer-Rao lower bound for the estimation variance of f_1 and ΔN from a dualband non-coherent signal with $N_1 = N_2 = 32$, SNR $= A_i^2/\sigma_w^2 = 0\,\mathrm{dB}$, variable gap ΔN and P exponentials with $A_i = 1$, $\phi_i = 0°$, $f_1 = 0.1$ and $f_i = f_{i-1} + \frac{2}{N_1+N_2}$

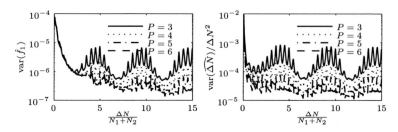

Figure 5.5: Cramer-Rao lower bound for the estimation variance of f_1 and ΔN from a dualband non-coherent signal with $N_1 = N_2 = 32$, SNR $= A_i^2/\sigma_w^2 = 0\,\mathrm{dB}$, variable gap ΔN and P exponentials with $A_i = 1$, $\phi_i = 0°$, $f_1 = 0.1$, $f_2 = f_1 + \frac{0.2}{N_1+N_2}$ and $f_i = f_{i-1} + \frac{2}{N_1+N_2}$ for $i > 2$

with a variable set of widely spaced sinusoids. A test signal is defined:

$$s_l[k_l; \boldsymbol{\theta}] = A_1 e^{j\phi_1} e^{j2\pi f_1 (N_c \pm \frac{\Delta N'}{2} + k_l)} + A_2 e^{j\phi_2} e^{j2\pi f_2 (N_c \pm \frac{\Delta N'}{2} + k_l)} +$$

$$+ \sum_{i=3}^{P} A_i e^{j\phi_i} e^{j2\pi (f_2 + (i-2)\Delta f)(N_c \pm \frac{\Delta N'}{2} + k_l)}.$$

Figure 5.5 shows the estimate variance of f_1 and ΔN. The range accuracy improves by increasing P, i. e. the lack of coherence in a dualband radar system can be partially compensated by estimating the band gap when an high enough number of widely spaced targets are present in the scenario. Figure 5.6 shows a comparison of the CRLB for the frequency estimate f_1 when the same signal is received by a coherent and a non-coherent system. The signal is composed by P exponentials, f_1 and f_2 are closely spaced, and the rest widely spaced. It is observed that the accuracy obtained for low window gaps, i. e. lower than $2 \cdot (N_1 + N_2)$, is the same for the non-coherent and the coherent signal reception. With higher gaps, however, the coherent approach shows a better accuracy.

5.2. Conclusions

The analysis of the CRLB for range estimation in dualband radar systems presented in this chapter shows an increase of the range accuracy for increasing band gap between the frequency bands. In coherent systems this improvement is observed for closely and widely spaced targets. In non-coherent systems, however, the accuracy improves only when a high enough number of widely spaced targets, i. e. with distances between them over the Rayleigh bound, are present in the signal. The presence of widely spaced targets permits to

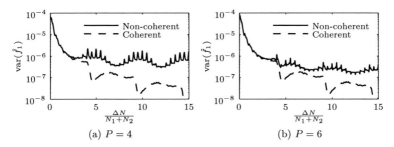

Figure 5.6: Cramer-Rao lower bound for the estimation variance of f_1 from a coherent and a non-coherent dualband signal with $N_1 = N_2 = 32$, SNR $= A_i^2/\sigma_w^2 = 0\,\text{dB}$, variable gap ΔN and P exponentials with $A_i = 1$, $\phi_i = 0°$, $f_1 = 0.1$, $f_2 = f_1 + \frac{0.2}{N_1+N_2}$ and $f_i = f_{i-1} + \frac{2}{N_1+N_2}$ for $i > 2$

estimate accurately enough the band gap, so that the range accuracy of the closely spaced targets also increases. A similar performance as in the coherent case is achieved. With increasing band gap, however, the number of required widely spaced targets to equal the performances of the non-coherent and the coherent systems becomes higher.

6. Range resolution in multiband radar systems

Several approaches to assess the resolution capability of a radar system have been analyzed in chapter 4. In this chapter the assessment of the range resolution of *coherent* and *non-coherent* dualband radar systems is addressed using the *hypothesis testing* or detection theory, as this approach is independent from the specific algorithm, presents a performance bound, and can be easily applied to the dualband case.

6.1. Hypothesis testing theory in dualband radar systems

The hypothesis testing theory is applied here to dualband radar signals. The hypotheses of none, one or two targets are tested with the received signal to decide how many targets are present in it. Depending on the assumed knowledge of the signal, different optimum detectors or discriminators can be defined. In the single-band case two types of *maximum likelihood* (ML) detectors were presented: based on the exact knowledge of the received signals and on their *probability density function* (PDF). In the dualband case, however, only the detector based on the exact knowledge of the PDF will be analyzed. In a non-coherent system the frequency distance between the two subbands is not exactly known and can be considered random. Therefore, the relation between the received signals at the two subbands is also unknown and they are considered independent to each other. For this reason, the assumption of exact knowledge of the received signal does not represent well the lack of coherence, as a knowledge of the received signals implies a knowledge of their relation. The assumption of exact knowledge of the PDF, however, represents well both coherent and non-coherent cases. The PDF depends on the autocorrelation matrix, where the lack of coherence is well represented by zeros at the lags where the subband data cross. Next, clairvoyant detectors for the coherent and non-coherent cases are presented.

Consider a radar signal sampled in the frequency domain at two different frequency bands $\mathcal{B}_i$:

$$x[n] = s[n] + w[n] \qquad n = \begin{cases} N_{10}, \ldots, N_{10} + N_1 - 1 & \text{at band } \mathcal{B}_1 \\ N_{20}, \ldots, N_{20} + N_2 - 1 & \text{at band } \mathcal{B}_2 \end{cases} \qquad (6.1)$$

where $w[n]$ is an additive *complex white Gaussian noise* (CWGN), $w[n] \sim \mathcal{CN}(0, \sigma_w^2)$, and $s[n]$ is a radar signal as described in chapter 2. The radar signal is decomposed as a sum of complex exponentials, each representing a single scattering center at a distance given by the normalized frequency f_i

$$s[n] = \sum_i h_i e^{j2\pi f_i n} = \sum_i A_i e^{j\phi_i} e^{j2\pi f_i n}. \tag{6.2}$$

This sequence can be split into two subwindow signals

$$s_{B_1}[k_1] = \sum_i h_{i,1} e^{j2\pi f_i k_1} = \sum_i A_i e^{j\phi_{i,1}} e^{j2\pi f_i k_1} \tag{6.3a}$$

$$s_{B_2}[k_2] = \sum_i h_{i,2} e^{j2\pi f_i k_2} = \sum_i A_i e^{j\phi_{i,2}} e^{j2\pi f_i k_2} \tag{6.3b}$$

with

$$k_1 = 0, \ldots, N_1 - 1$$
$$k_2 = 0, \ldots, N_2 - 1$$

and

$$h_{i,1} = h_i e^{j2\pi f_i N_{10}} \qquad\qquad \phi_{i,1} = \phi_i + 2\pi f_i N_{10}$$
$$h_{i,2} = h_i e^{j2\pi f_i N_{20}} \qquad\qquad \phi_{i,2} = \phi_i + 2\pi f_i N_{20}$$

In a coherent system, the relative distance between the two subbands is exactly known, and therefore, there is a known relation between the amplitude phases at the two subbands

$$\frac{h_{i,2}}{h_{i,1}} = e^{j2\pi f_i (N_{20} - N_{10})} \tag{6.4a}$$

$$\phi_{i,2} - \phi_{i,1} = 2\pi f_i (N_{20} - N_{10}). \tag{6.4b}$$

In a non-coherent system, the relative distance between the frequency subbands is only known approximately, and the phases $\phi_{i,1}$ and $\phi_{i,2}$ are considered independent to each other.

The considered hypotheses for the radar analysis are:

$$x[n] = s_0[n] + w[n] \qquad\qquad \mathcal{H}_0 \text{ no target} \tag{6.5a}$$
$$x[n] = s_1[n] + w[n] \qquad\qquad \mathcal{H}_1 \text{ one target} \tag{6.5b}$$
$$x[n] = s_2[n] + w[n] \qquad\qquad \mathcal{H}_2 \text{ two targets} \tag{6.5c}$$

with

$$s_0[n] = 0$$
$$s_1[n] = h_m e^{j2\pi f_m n} = A_m e^{j\phi_m} e^{j2\pi f_m n}$$
$$s_2[n] = h_1 e^{j2\pi f_1 n} + h_2 e^{j2\pi f_2 n} = A_1 e^{j\phi_1} e^{j2\pi f_1 n} + A_2 e^{j\phi_2} e^{j2\pi f_2 n}.$$

The sequences can also be represented by vectors

$$\mathbf{x} = \mathbf{s}_0 + \mathbf{w} \qquad \mathcal{H}_0 \text{ no target} \qquad (6.6a)$$

$$\mathbf{x} = \mathbf{s}_1 + \mathbf{w} \qquad \mathcal{H}_1 \text{ one target} \qquad (6.6b)$$

$$\mathbf{x} = \mathbf{s}_2 + \mathbf{w} \qquad \mathcal{H}_2 \text{ two targets} \qquad (6.6c)$$

with

$$\mathbf{x} = \left(x[N_{10}] \quad \ldots \quad x[N_{10} + N_1 - 1] \quad x[N_{20}] \quad \ldots \quad x[N_{20} + N_2 - 1] \right)^T$$

$$\mathbf{s}_i = \left(s_i[N_{10}] \quad \ldots \quad s_i[N_{10} + N_1 - 1] \quad s_i[N_{20}] \quad \ldots \quad s_i[N_{20} + N_2 - 1] \right)^T$$

$$\mathbf{w} = \left(w[N_{10}] \quad \ldots \quad w[N_{10} + N_1 - 1] \quad w[N_{20}] \quad \ldots \quad w[N_{20} + N_2 - 1] \right)^T .$$

The complex amplitudes of the received signals are considered independent random complex Gaussian variables, i.e. $h_1 \sim \mathcal{CN}(0, \sigma^2)$, $h_2 \sim \mathcal{CN}(0, \sigma^2)$ and $h_m \sim \mathcal{CN}(0, \sigma_m^2)$, so that the received radar signals have also a complex Gaussian distribution $\mathbf{x} \sim \mathcal{CN}(0, \mathbf{C}_{s_i} + \sigma_w^2 \mathbf{I})$, where $\mathbf{C}_{s_i}$ is the autocorrelation matrix of the signal $s_i[n]$—corresponding to hypothesis $\mathcal{H}_i$—defined as

$$\mathbf{C}_{s_i} = E(\mathbf{s}_i \cdot \mathbf{s}_i^H). \qquad (6.7)$$

Applying the ML criteria as in the single-band case, to minimize the *probability of error* P_e the selected hypothesis will maximize the test function

$$T_i(\mathbf{x}) = -\ln\left(\det\left(\mathbf{C}_{s_i} + \sigma_w^2 \mathbf{I}\right)\right) - \mathbf{x}^H \left(\mathbf{C}_{s_i} + \sigma_w^2 \mathbf{I}\right)^{-1} \mathbf{x}. \qquad (6.8)$$

The main difference between the coherent and the non-coherent detector lies on the dependence between the subband vectors. In the coherent case, as the band gap is known, there is a known relation between the samples at window $\mathcal{B}_1$ and at $\mathcal{B}_2$, and therefore the matrix $\mathbf{C}_{s_i}$ have in general no zero elements. In the non-coherent case, however, there is an unknown phase shift between the oscillations corresponding to a scatterer at each subbands, so that their are considered independent to each other, and the matrix $\mathbf{C}_{s_i}$ will have zeros where data of different subbands cross.

6.1.1. Coherent system

As in the single-band case, the vector $\mathbf{s}_i$ is expressed as the product of a constant matrix $\mathbf{H}_i$ composed by deterministic oscillation sequences at frequencies

f_i, and a vector $\boldsymbol{\theta}_i$ composed by their random complex amplitudes h_i

$$\mathbf{s}_1 = \mathbf{H}_2 \cdot \boldsymbol{\theta}_1 = \begin{pmatrix} 1 \\ e^{j\omega_m} \\ \vdots \\ e^{j\omega_m(N_1-1)} \\ e^{j\omega_m \Delta N'} \\ e^{j\omega_m(\Delta N'+1)} \\ \vdots \\ e^{j\omega_m(\Delta N'+N_2-1)} \end{pmatrix} \cdot \left(h_m \right) \tag{6.9a}$$

$$\mathbf{s}_2 = \mathbf{H}_2 \cdot \boldsymbol{\theta}_2 = \begin{pmatrix} 1 & 1 \\ e^{j\omega_1} & e^{j\omega_2} \\ \vdots & \vdots \\ e^{j\omega_1(N_1-1)} & e^{j\omega_2(N-1)} \\ e^{j\omega_1 \Delta N'} & e^{j\omega_2 \Delta N'} \\ e^{j\omega_1(\Delta N'+1)} & e^{j\omega_1(\Delta N'+1)} \\ \vdots & \vdots \\ e^{j\omega_1(\Delta N'+N_2-1)} & e^{j\omega_1(\Delta N'+N_2-1)} \end{pmatrix} \cdot \begin{pmatrix} h_1 \\ h_2 \end{pmatrix} \tag{6.9b}$$

where $\Delta N' = N_{20} - N_{10}$, so that

$$\mathbf{C}_{s_i} = \mathbf{H}_i \mathbf{C}_{\theta_i} \mathbf{H}_i^H.$$

Taking into account that

$$\mathbf{C}_{\theta_1} = \sigma_m^2 \tag{6.10a}$$

$$\mathbf{C}_{\theta_2} = \sigma^2 \mathbf{I}_2 \tag{6.10b}$$

the test function can be reduced to

$$T_1(\mathbf{x}) = -\ln\left(\frac{\sigma_m^2}{\sigma_w^2} \mathbf{H}_1^H \mathbf{H}_1 + 1 \right)$$

$$- \frac{1}{\sigma_w^2} \mathbf{x}^H \left(\mathbf{I}_N - \frac{\mathbf{H}_1 \mathbf{H}_1^H}{\frac{\sigma_w^2}{\sigma_m^2} + \mathbf{H}_1^H \mathbf{H}_1} \right) \mathbf{x} \tag{6.11a}$$

$$T_2(\mathbf{x}) = -\ln\left(\det\left(\frac{\sigma^2}{\sigma_w^2} \mathbf{H}_2^H \mathbf{H}_2 + \mathbf{I}_2 \right) \right)$$

$$- \frac{1}{\sigma_w^2} \mathbf{x}^H \left(\mathbf{I}_N - \mathbf{H}_2 \left(\frac{\sigma_w^2}{\sigma^2} \mathbf{I}_2 + \mathbf{H}_2^H \mathbf{H}_2 \right)^{-1} \mathbf{H}_2^H \right) \mathbf{x} \tag{6.11b}$$

where common terms have been suppressed, and the determinant property

$$\det\left(\mathbf{A}\mathbf{A}^H + \mathbf{I}\right) = \det\left(\mathbf{A}^H\mathbf{A} + \mathbf{I}\right)$$

and the matrix inversion lemma

$$(\mathbf{A} + \mathbf{U}\mathbf{C}\mathbf{V})^{-1} = \mathbf{A}^{-1} - \mathbf{A}^{-1}\mathbf{U}(\mathbf{C}^{-1} + \mathbf{V}\mathbf{A}^{-1}\mathbf{U})^{-1}\mathbf{V}\mathbf{A}^{-1}$$

have been used.

As there is no analytical expression for the *probability of resolution* P_R, defined in terms of hypothesis testing as

$$P_R = P(\mathcal{H}_2 | \mathcal{H}_2), \tag{6.12}$$

a Monte Carlo analysis is necessary. The following *autocorrelation sequence* (ACS) for the $\mathcal{H}_2$ hypothesis is considered

$$r_{s_2 s_2}[m] = \sigma^2 e^{j2\pi f_1 m} + \sigma^2 e^{j2\pi f_2 m}$$

corresponding to a signal with two tones with independent complex Gaussian amplitudes with equal variance σ^2. The signal-to-noise ratio is defined as SNR $= \sigma^2/\sigma_w^2$. For the one-target signal, the following one-tone ACS

$$r_{s_1 s_1}[m] = \sigma_m^2 e^{j2\pi f_m m}$$

is optimized over f_m and σ_m^2 to minimize the difference—in a quadratic error sense—to $r_{s_2 s_2}[m]$. The one-target ACS which is closer to the actual two-target ACS is selected to obtain a bound for P_R. The random two-target signal is evaluated with the test corresponding to the sequence which—in mean sense—minimizes the difference between the hypothesis sequences.

Figure 6.1 shows the resolution capability of the clairvoyant dualband coherent detector for different frequency distances $\Delta f = f_2 - f_1$. A total number of 10^4 trials per band gap point has been realized. As test signal, two complex oscillations with independent random complex Gaussian amplitude are used. As in the single-band case, the behavior is independent of the center frequency of the sequences, so that only one f_1 case per Δf is shown. It is also observed an oscillating behavior of the P_R with period $1/\Delta f$. This is consistent with the periodicity of the envelope of a two-target signal, which can be expressed as:

$$s_2[n] = e^{j(\omega_1 n + \phi_1)} + e^{j(\omega_2 n + \phi_2)} =$$
$$= 2e^{j\left(\frac{\omega_1 + \omega_2}{2}n + \frac{\phi_1 + \phi_2}{2}\right)} \cos\left(\frac{\omega_1 - \omega_2}{2}n + \frac{\phi_1 - \phi_2}{2}\right). \tag{6.13}$$

The envelope of the complex oscillation has a period of $2/\Delta f$, two times the periodicity observed in the resolution performance. This periodic effect was also observed in the range accuracy analysis for dualband systems.

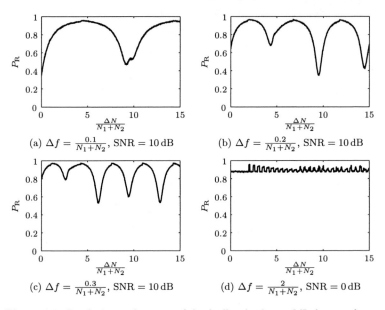

Figure 6.1: Resolution performance of the dualband coherent ML detector for random radar signals with known Gaussian PDF, $N_1 = N_2 = 32$, $f_1 = 0.1$, $f_2 = f_1 + \Delta f$ and variable gap ΔN

6.1.2. Non-coherent system

The vector $\mathbf{s}_i$ is expressed as the product of a constant matrix $\mathbf{H}_i$ composed by deterministic oscillation sequences at frequencies f_i, and a vector $\boldsymbol{\theta}_i$ composed by their random complex amplitudes h_i

$$\mathbf{s}_1 = \begin{pmatrix} 1 & 0 \\ e^{j\omega_m} & 0 \\ \vdots & \vdots \\ e^{j\omega_m(N_1-1)} & 0 \\ 0 & 1 \\ 0 & e^{j\omega_m} \\ \vdots & \vdots \\ 0 & e^{j\omega_m(N_2-1)} \end{pmatrix} \cdot \begin{pmatrix} h_{m,1} \\ h_{m,2} \end{pmatrix} \tag{6.14a}$$

$$
\mathbf{s}_2 =
\begin{pmatrix}
1 & 1 & 0 & 0 \\
e^{j\omega_1} & e^{j\omega_2} & 0 & 0 \\
\vdots & \vdots & \vdots & \vdots \\
e^{j\omega_1(N_1-1)} & e^{j\omega_2(N_1-1)} & 0 & 0 \\
0 & 0 & 1 & 1 \\
0 & 0 & e^{j\omega_1} & e^{j\omega_2} \\
\vdots & \vdots & \vdots & \vdots \\
0 & 0 & e^{j\omega_1(N_2-1)} & e^{j\omega_2(N_2-1)}
\end{pmatrix}
\cdot
\begin{pmatrix}
h_{1,1} \\
h_{1,2} \\
h_{2,1} \\
h_{2,2}
\end{pmatrix}
\tag{6.14b}
$$

so that

$$
\mathbf{C}_{s_i} = \mathbf{H}_i \mathbf{C}_{\theta_i} \mathbf{H}_i^H .
$$

Taking into account that

$$
\mathbf{C}_{\theta_1} = \sigma_m^2 \mathbf{I}_2 \tag{6.15a}
$$

$$
\mathbf{C}_{\theta_2} = \sigma^2 \mathbf{I}_4 \tag{6.15b}
$$

the test function can be reduced to

$$
T_1(\mathbf{x}) = -\ln\left(\det\left(\frac{\sigma_m^2}{\sigma_w^2}\mathbf{H}_1^H\mathbf{H}_1 + \mathbf{I}_2\right)\right)
$$
$$
- \frac{1}{\sigma_w^2}\mathbf{x}^H\left(\mathbf{I}_N - \mathbf{H}_1\left(\frac{\sigma_w^2}{\sigma_m^2}\mathbf{I}_2 + \mathbf{H}_1^H\mathbf{H}_1\right)^{-1}\mathbf{H}_1^H\right)\mathbf{x} \tag{6.16a}
$$

$$
T_2(\mathbf{x}) = -\ln\left(\det\left(\frac{\sigma^2}{\sigma_w^2}\mathbf{H}_2^H\mathbf{H}_2 + \mathbf{I}_4\right)\right)
$$
$$
- \frac{1}{\sigma_w^2}\mathbf{x}^H\left(\mathbf{I}_N - \mathbf{H}_2\left(\frac{\sigma_w^2}{\sigma^2}\mathbf{I}_4 + \mathbf{H}_2^H\mathbf{H}_2\right)^{-1}\mathbf{H}_2^H\right)\mathbf{x} \tag{6.16b}
$$

where common terms have been suppressed and the same determinant properties and lemmas as in the coherent case are used.

A Monte Carlo analysis is realized to obtain the P_{R}. The considered autocorrelation sequence for the $\mathcal{H}_2$ hypothesis is

$$
r_{s_2 s_2}[m] = \sigma^2 e^{j2\pi f_1 m} + \sigma^2 e^{j2\pi f_2 m}
$$

for the samples corresponding to the same observation window. The correlation between samples from different observation window is zero because of incoherency. For the one-target signal, the following one-tone ACS

$$
r_{s_1 s_1}[m] = \sigma_m^2 e^{j2\pi f_m m}
$$

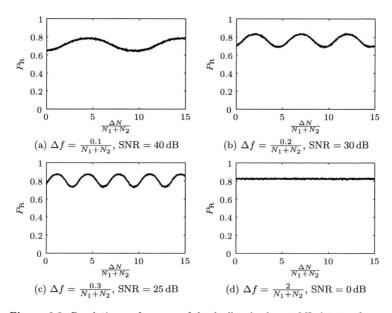

Figure 6.2: Resolution performance of the dualband coherent ML detector for random radar signals with known PDF. $N_1 = N_2 = 32$, $f_1 = 0.1$, $f_2 = f_1 + \Delta f$ and variable gap ΔN

is optimized—as in the coherent case—over f_m and σ_m^2 to minimize the difference to $r_{s_2 s_2}[m]$. The signal to noise ratio is defined as SNR $= \sigma^2/\sigma_w^2$.

Figure 6.2 shows the resolution capability of the clairvoyant dualband non-coherent detector for different frequency distances $\Delta f = f_2 - f_1$. A total number of 10^4 trials per band gap point has been realized. As test signal, two complex oscillations with independent random complex Gaussian amplitude are used. Please note that there is a relation between the amplitudes at the two observation windows, as only one amplitude per tone is generated, but the detector makes no use of this information. As in the previous case, the behavior is independent of the center frequency of the sequences, so that only one f_1 case per Δf is shown. As in the coherent case, an oscillating behavior of the P_R with period $1/\Delta f$ is observed. In the non-coherent case, however, the amplitude of this oscillation effect is lower. For $\Delta f = 2/(N_1 + N_2)$ the P_R can be considered independent of the band gap ΔN.

6.2. Conclusions

Range resolution in dualband coherent and non-coherent radar systems have been analyzed with the hypothesis testing theory. The analysis shows an oscillating effect in the probability of resolution P_R as a function of the frequency band gap for closely spaced targets. The period of this oscillation is inversely proportional to the distance between the two targets present in the signal. This effect is less defined in the non-coherent case, where the oscillations in the P_R are present, but with lower intensity. For distances over the Rayleigh bound, the resolution capability is independent from the band gap. The used test functions are the so-called clairvoyant detectors. They assume an unavailable knowledge of some signal parameters and show an upper bound in the resolution performance. The resolution in specific radar systems typically exceeds the clairvoyant detector resolution limit, although their qualitative behaviors would be similar. It can be concluded that no general improvement of the resolution capability in a dualband radar system is expected with increasing band gap.

7. Spectral estimation for multiband radar systems

In a multiband radar system the radar scenario is observed at different frequency bands. For sensor data collected in the frequency domain, the data set is split into more than one observation window. If the relative distance between the windows is exactly known, so that the phase continuity of an exponential is assured, the data set is said to be *coherent*. If not, if no phase relation can be established, the data set is *non-coherent*. In this chapter some of the spectral estimation techniques presented in chapter 3 are modified to address the multi-window case. Depending on how the algorithms relate the samples of the different observation windows to each other, coherent or non-coherent approaches—algorithms, which process coherent or non-coherent data sets, respectively—can be defined.

The general case of a signal observed at different windows $\mathcal{B}_l$ is reduced here to the dualband case, where the sensor data is collected at two windows. The results obtained in the dualband case, however, can be extended to the multiband one. A general radar signal as described in chapter 2 is assumed for all the analysis in this chapter

$$s[n] = \sum_{i=1}^{P} h_i z_i^n \tag{7.1}$$

where the poles $z_i = \rho_i e^{j2\pi f_i}$ represent targets at normalized distances f_i and with frequency decays given by ρ_i. Please note that as the data is sampled in the frequency domain, frequency is represented by the sequence index n. In some cases no decay with frequency is assumed for the radar targets so that $z_i = e^{j2\pi f_i}$. The signal is buried in additive *complex white Gaussian noise* (CWGN), $w[n] \sim \mathcal{CN}(0, \sigma_w^2)$

$$x[n] = s[n] + w[n] \tag{7.2}$$

and the sequence index is expressed as

$$n = \begin{cases} N_{10}, \ldots, N_{10} + N_1 - 1 & \text{at } \mathcal{B}_1 \\ N_{20}, \ldots, N_{20} + N_2 - 1 & \text{at } \mathcal{B}_2 \end{cases} \tag{7.3}$$

Next, classical spectral estimators are discussed, followed by autoregressive and eigenanalysis based techniques. Coherent and non-coherent approaches are identified for each group.

7.1. Classical spectral estimation

Classical spectral algorithms are based on the *discrete Fourier transform* (DFT) of data vectors, e. g. periodogram, or *autocorrelation sequence* (ACS) estimates, e. g. correlogram. Coherent and non-coherent approaches can be found for both groups.

7.1.1. Non-coherent dualband

In non-coherent approaches the relation between the samples of the two subwindows is not used, i. e. data vectors with samples of only one subband, or ACS estimates with maximum autocorrelation lag L smaller than one subwindow length N_l, are transformed by the DFT.

Approaches as Bartlett or Welch can be directly applied. Periodograms of segments from the different subbands are averaged to reduce the estimate variance. The resolution, as in the single-band case, is proportional to $1/L$, where L is the length of the data segments. The gain from the dualband scenario is the increase of data samples, which allows the selection of a greater L, i. e. increase in the resolution, by maintaining the estimate variance. With respect to an equivalent single band case with $N = N_1 + N_2$ no gain can be achieved.

Correlogram approaches are also applied directly to ACS estimates obtained with samples of all the observation windows, e. g.

$$\hat{r}_{xx}[m] = \frac{1}{N_1 + N_2 - 2m} \left(\sum_{n=N_{10}}^{N_{10}+N_1-m-1} x[n+m]x^*[n] + \sum_{n=N_{20}}^{N_{20}+N_2-m-1} x[n+m]x^*[n] \right). \quad (7.4)$$

The resolution is proportional to $1/L$, where L is the maximum autocorrelation lag. The gain from the dualband scenario is, as in the periodogram case, the possibility to choose a greater L by maintaining the estimate variance. However, the same results are obtained as for a single-band case with $N = N_1 + N_2$.

7.1.2. Coherent dualband

In a coherent approach, the relation between samples of the two subbands is used. Therefore, the DFT process data vectors or ACS estimates with zeros in between, as the information of the band gap is missing. Of course, as the relation between the subbands is used, coherence is required.

A detailed analysis of the effects of the band gap, coherence errors and calibration errors on the spectral estimate has been shown by Siart in [75].

Here only some final results are given. The periodogram is selected for the analysis. The obtained results, however, are valid for other approaches.

Figure 7.1 shows the periodogram of two rectangular windows as function of the gap ΔN between them. With increasing band gap the main lobe of the transformation becomes narrower, which means better resolution and accuracy. However, the secondary lobe becomes becomes greater and comes near to the main one. This leads to ambiguities, as only peaks with a separation between them lower than the distance between the lobes can be unambiguously assigned to targets.

7.1.3. Conclusions

Two main approaches to apply classical spectral analysis to dualband signals have been shown here: non-coherent approaches, where no information of the band gap is required, and coherent approaches, where the band gap must be exactly known.

The dualband coherent approaches show a narrower main lobe, i. e. a better resolution and accuracy can be expected, with increasing gap. However, ambiguity problems arise and become acuter also with increasing window gap.

The dualband non-coherent approaches do not show ambiguity problems and they perform as the corresponding single-band ones with equal total number of samples. Exception as this point is the periodogram—method which also shows the best resolution from the classical algorithms—as it is based on the Fourier transformation of the whole data set, which becomes impossible with the non-coherent approach. A minimum of two independent windows are required.

7.2. Autoregressive spectral estimation

The autoregressive methods presented in section 3.2 are now applied to signals observed at different windows. The analysis is focused on the modified covariance method. The results and conclusions, however, are also valid for other AR techniques like the covariance or the autocorrelation approaches. Recalling from section 3.2, in the modified covariance method the AR coefficients were found by minimizing the forward and backward linear prediction errors simultaneously. The set of filter coefficients was given by eq. (3.21), where the involved matrices and vectors were build using as basis the so-called *data matrix* of order L,

$$
\mathbf{X}_L = \begin{pmatrix} x[L] & \cdots & x[0] \\ \vdots & \ddots & \vdots \\ x[N-1] & \cdots & x[N-L-1] \end{pmatrix}
$$

built with the data samples $x[n]$ for $n = 0, \ldots, N-1$.

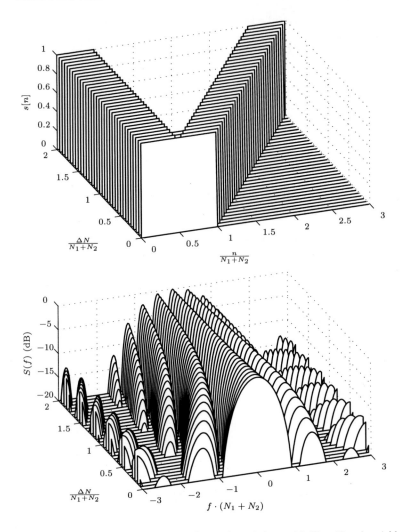

Figure 7.1: DFT of two rectangular observation windows with $N_l = 32$ and variable gap ΔN between them

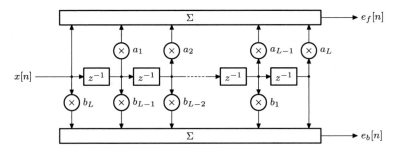

Figure 7.2: Linear prediction error filter for dualband non-coherent processing. The same filter structure as in the single-band case is preserved.

To include the information of the two observation windows new data matrices are defined. Two general approaches to construct a dualband data matrix are presented, resulting in a coherent and a non-coherent approach.

7.2.1. Non-coherent dualband AR

The non-coherent data matrix $\mathbf{X}_{\mathrm{nc}}$ is built as the vertical superposition of the subband matrices $\mathbf{X}_{L,1}$ and $\mathbf{X}_{L,2}$

$$\mathbf{X}_{\mathrm{nc}} = \begin{pmatrix} \mathbf{X}_{L,1} \\ \mathbf{X}_{L,2} \end{pmatrix} \tag{7.5}$$

where

$$\mathbf{X}_{L,1} = \begin{pmatrix} x[N_{10} + L] & \cdots & x[N_{10}] \\ \vdots & \ddots & \vdots \\ x[N_{10} + N_1 - 1] & \cdots & x[N_{10} + N_1 - L - 1] \end{pmatrix}$$

$$\mathbf{X}_{L,2} = \begin{pmatrix} x[N_{20} + L] & \cdots & x[N_{20}] \\ \vdots & \ddots & \vdots \\ x[N_{20} + N_2 - 1] & \cdots & x[N_{20} + N_2 - L - 1] \end{pmatrix}.$$

Following the same development as in the single-band case, the expression for the forward $e_f[n]$ and backward $e_b[n]$ prediction errors of FIR predictors with L coefficients—as shown in figure 7.2—is expressed in vector form as

$$\mathbf{e} = \begin{pmatrix} \mathbf{e}_f \\ \mathbf{e}_b^* \end{pmatrix} = \begin{pmatrix} \mathbf{X}_{\mathrm{nc}} \\ \mathbf{X}_{\mathrm{nc}}^* \mathbf{J} \end{pmatrix} \begin{pmatrix} 1 \\ \mathbf{a} \end{pmatrix} \tag{7.6}$$

where the vectors are defined as

$$
\mathbf{e}_f = \begin{pmatrix} e_f[N_{10} + L] \\ \vdots \\ e_f[N_{10} + N_1 - 1] \\ e_f[N_{20} + L] \\ \vdots \\ e_f[N_{20} + N_2 - 1] \end{pmatrix} \quad \mathbf{e}_b = \begin{pmatrix} e_b[N_{10}] \\ \vdots \\ e_b[N_{10} + N_1 - L - 1] \\ e_b[N_{20}] \\ \vdots \\ e_b[N_{20} + N_2 - L - 1] \end{pmatrix} \quad \mathbf{a} = \begin{pmatrix} a_1 \\ \vdots \\ a_L \end{pmatrix}
$$

and the following relation between the forward a_l and backward b_l filter coefficients—defined in equations (3.12) and (3.15)—in an AR process has been used

$$
b_l = a_l^*. \tag{7.7}
$$

The optimum set of filter coefficients $\mathbf{a}$ is found by minimizing the energy of the prediction error $\mathbf{e}^H \mathbf{e}$ and results in

$$
\mathbf{a} = -\left(\mathbf{X}^H \mathbf{X} \right)^{-1} \mathbf{X}^H \mathbf{x} \tag{7.8}
$$

where the vector $\mathbf{x}$ and the matrix $\mathbf{X}$ are built, respectively, with the first row and with the rows 2 to $L+1$ of the matrix $\begin{pmatrix} \mathbf{X}_L \\ \mathbf{X}_L^* \mathbf{J} \end{pmatrix}$

$$
\mathbf{x} = \begin{pmatrix} x[N_{10} + L] \\ \vdots \\ x[N_{10} + N_1 - 1] \\ x[N_{20} + L] \\ \vdots \\ x[N_{20} + N_2 - 1] \\ x^*[N_{10}] \\ \vdots \\ x^*[N_{10} + N_1 - L - 1] \\ x^*[N_{20}] \\ \vdots \\ x^*[N_{20} + N_2 - L - 1] \end{pmatrix}
$$

$$\mathbf{X} = \begin{pmatrix} x[N_{10} + L - 1] & \cdots & x[N_{10}] \\ \vdots & \ddots & \vdots \\ x[N_{10} + N_1 - 2] & \cdots & x[N_{10} + N_1 - L - 1] \\ x[N_{20} + L - 1] & \cdots & x[N_{20}] \\ \vdots & \ddots & \vdots \\ x[N_{20} + N_2 - 2] & \cdots & x[N_{20} + N_2 - L - 1] \\ x^*[N_{10} + 1] & \cdots & x^*[N_{10} + L] \\ \vdots & \ddots & \vdots \\ x^*[N_{10} + N_1 - L] & \cdots & x^*[N_{10} + N_1 - 1] \\ x^*[N_{20} + 1] & \cdots & x^*[N_{20} + L] \\ \vdots & \ddots & \vdots \\ x^*[N_{20} + N_2 - L] & \cdots & x^*[N_{20} + N_2 - 1] \end{pmatrix} .$$

This structure corresponds to a predictor filter as shown in figure 7.2, which coincides with the filter used in the single-band case. It is noted that no information about the relative position between the two bands is used.

Once the filter coefficients are obtained, the estimate of the sequence spectrum is

$$P_{\mathrm{AR}}(f) = \sigma_e^2 \cdot |H_{\mathrm{AR}}(e^{j2\pi f})|^2 = \frac{\sigma_e^2}{|A(e^{j2\pi f})|^2} \tag{7.9}$$

with

$$A(z) = 1 + a_1 z^{-1} + \cdots + a_L z^{-L}$$

and σ_e^2 the variance of the error. The poles of the spectrum, which some of them correspond to the radar targets contained in the signal, are the zeros of the polynomial $A(z)$.

This approach has the advantage that an exact knowledge of the relative position of the two observation windows is not required because the information of the gap is not used. Due to the structure of the data matrix, the performance of the estimator is expected to be the same as for the single-band algorithm with $N_1 + N_2$ data samples. Therefore, this non-coherent approach results in a robust way to use non-coherent data sets for spectrum estimation, and achieves the same performance as the single-window algorithm with an equivalent data set length equal to the sum of the subwindows lengths. As drawback, no advantage of the gap information can be obtained.

7.2.2. Coherent dualband AR

In this approach, the relation between the samples from the two observation windows is explicitly used in the predictor filter, which is represented in the

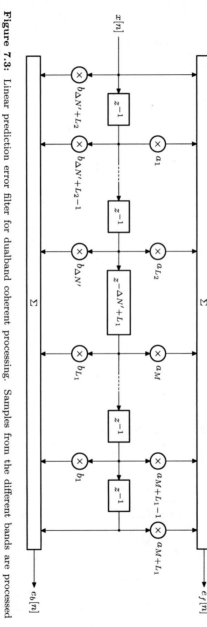

Figure 7.3: Linear prediction error filter for dualband coherent processing. Samples from the different bands are processed coherently.

figure 7.3. The data matrix $\mathbf{X}_c$ is built as the horizontal juxtaposition of the subband matrices $\mathbf{X}_{L_1,1}$ and $\mathbf{X}_{L_2,2}$

$$\mathbf{X}_c = \begin{pmatrix} \mathbf{X}_{L_2,2} & \mathbf{X}_{L_1,1} \end{pmatrix} \tag{7.10}$$

where

$$\mathbf{X}_{L_1,1} = \begin{pmatrix} x[N_{10} + L_1] & \cdots & x[N_{10}] \\ \vdots & \ddots & \vdots \\ x[N_{10} + N_1 - 1] & \cdots & x[N_{10} + N_1 - L_1 - 1] \end{pmatrix}$$

$$\mathbf{X}_{L_2,2} = \begin{pmatrix} x[N_{20} + L_2] & \cdots & x[N_{20}] \\ \vdots & \ddots & \vdots \\ x[N_{20} + N_2 - 1] & \cdots & x[N_{20} + N_2 - L_2 - 1] \end{pmatrix}$$

and the following relation has been assumed in order to use all the available data and to obtain the same number of rows in both matrices

$$N_2 - N_1 = L_2 - L_1. \tag{7.11}$$

Following the same development as in the non-coherent approach, the expression for the linear prediction error results in

$$\mathbf{e} = \begin{pmatrix} \mathbf{e}_f \\ \mathbf{e}_b^* \end{pmatrix} = \begin{pmatrix} \mathbf{X}_c \\ \mathbf{X}_c^* \mathbf{J} \end{pmatrix} \begin{pmatrix} 1 \\ \mathbf{a} \end{pmatrix} \tag{7.12}$$

where

$$\mathbf{e}_f = \begin{pmatrix} e_f[N_{20} + L] \\ \vdots \\ e_f[N_{20} + N_2 - 1] \end{pmatrix} \quad \mathbf{e}_b = \begin{pmatrix} e_b[N_{10}] \\ \vdots \\ e_b[N_{10} + N_1 - L - 1] \end{pmatrix} \quad \mathbf{a} = \begin{pmatrix} a_1 \\ \vdots \\ a_{L_2} \\ a_M \\ \vdots \\ a_{M+L_1} \end{pmatrix}$$

with

$$M = N_{20} - N_{10} + N_2 - N_1 \tag{7.13}$$

and where the following relation between forward and backward predictor filter coefficients has been included

$$b_l = a_l^*. \tag{7.14}$$

The optimum set of filter coefficients $\mathbf{a}$ is found by minimizing the energy of the prediction error $\mathbf{e}^H \mathbf{e}$ and results in

$$\mathbf{a} = -\left(\mathbf{X}^H \mathbf{X}\right)^{-1} \mathbf{X}^H \mathbf{x} \tag{7.15}$$

where the vector $\mathbf{x}$ and the matrix $\mathbf{X}$ are built—as in the non-coherent case—with the first row and with the rows 2 to $L+1$ of the matrix $\begin{pmatrix} \mathbf{X}_L \\ \mathbf{X}_L^* \mathbf{J} \end{pmatrix}$, respectively

$$
\mathbf{x} = \begin{pmatrix}
x[N_{20} + L_2] \\
\vdots \\
x[N_{20} + N_2 - 1] \\
x^*[N_{10}] \\
\vdots \\
x^*[N_{10} + N_1 - L_1 - 1]
\end{pmatrix}
$$

$$
\mathbf{X} = \begin{pmatrix}
x[N_{20} + L_2 - 1] & \cdots & x[N_{10}] \\
\vdots & \ddots & \vdots \\
x[N_{20} + N_2 - 2] & \cdots & x[N_{10} + N_1 - L_1 - 1] \\
x^*[N_{10} + 1] & \cdots & x^*[N_{20} + L_2] \\
\vdots & \ddots & \vdots \\
x^*[N_{10} + N_1 - L_1] & \cdots & x^*[N_{20} + N_2 - 1]
\end{pmatrix}.
$$

Once the filter coefficients are obtained, the estimate of the sequence spectrum is

$$
P_{\mathrm{AR}}(f) = \sigma_e^2 \cdot |H_{\mathrm{AR}}(e^{j2\pi f})|^2 = \frac{\sigma_e^2}{|A(e^{j2\pi f})|^2} \tag{7.16}
$$

with

$$
A(z) = 1 + a_1 z^{-1} + \cdots + a_{L_2} z^{-L_2} + a_M z^{-M} + \cdots a_{M+L_1} z^{-M-L_1}
$$

and σ_e^2 the variance of the error. The poles of the signal are the zeros of the polynomial $A(z)$.

This system of equations corresponds to a predictor filter with the structure shown in figure 7.3. The first part of the filter—with coefficients $a_1, \ldots, a_{L_2}$—uses for the prediction of a sample at window $\mathcal{B}_l$ signal samples of the same observation window. The second part of the filter—coefficients $a_M, \ldots, a_{M+L_1}$—uses samples of the other window, so that the relation between samples at different windows is used. This is the main difference with the non-coherent method, the gap information is used and appears in the filter structure.

Of course, as the gap information is required, it must be known exactly, i.e. coherence between the data must be assured. The main drawback of the method is the linear relation between the order of the polynomial $A(z)$ and the gap. The order is now $M + L_1$ and this results in the appearance of a higher number of zeros which difficult the selection of the ones corresponding to the signal. This may be solved by combining the coherent with the non-coherent approaches, as shown in section 7.4.

7.2.3. Conclusions

Two main approaches to apply autoregressive spectral analysis to dualband signals have been shown: non-coherent approaches, where no information of the band gap is required, and coherent approaches, where the band gap must be exactly known.

The non-coherent approach is a robust way to use multiband data sets and achieves the same performance as the single-band algorithm with an equivalent data set length equal to the sum of the subband lengths. As drawback, no advantage of the gap information can be obtained.

In the coherent approach, the relation between data samples of different subbands is explicitly used and appears in the data matrix structure. Its main drawback is the linear increase of possible targets with increasing band gap, which complicates the selection of the targets corresponding to the received signal. This may be solved by combining the coherent with the non-coherent approaches, as shown in section 7.4.

Figure 7.4 shows different AR range profile estimates applied to the frequency response of a two-cylinder scenario obtained with the UTD approximation as described in appendix A. The estimates obtained by processing the subbands independently do not resolve the two cylinders, the presence of two targets can be guess, but they are not clearly resolved. In the dualband estimates the two cylinders can be identified. The coherent dualband approach shows the two targets explicitly, but also a ripple along the whole range. This problem is caused by the high-order polynomial involved in the coherent dualband approach. This undesired ripple disturbs the estimate and can origin ambiguity problems.

7.3. MUSIC spectral estimation

The MUSIC algorithms presented in chapter 3 are now applied to signals observed at different windows. Recalling from section 3.3, to obtain the signal poles from the received noisy signal, an eigenanalysis was done to the autocorrelation-like matrices

$$
\mathbf{X}_L^H \mathbf{X}_L \quad \text{or} \quad \begin{pmatrix} \mathbf{X}_L \\ \mathbf{X}_L^* \mathbf{J} \end{pmatrix}^H \begin{pmatrix} \mathbf{X}_L \\ \mathbf{X}_L^* \mathbf{J} \end{pmatrix}
$$

where the matrix $\mathbf{X}_L$ is the data matrix of order L used in the AR algorithms

$$
\mathbf{X}_L = \begin{pmatrix} x[L] & \cdots & x[0] \\ \vdots & \ddots & \vdots \\ x[N-1] & \cdots & x[N-L-1] \end{pmatrix}
$$

and built with the data samples $x[n]$ for $n = 0, \ldots, N-1$.

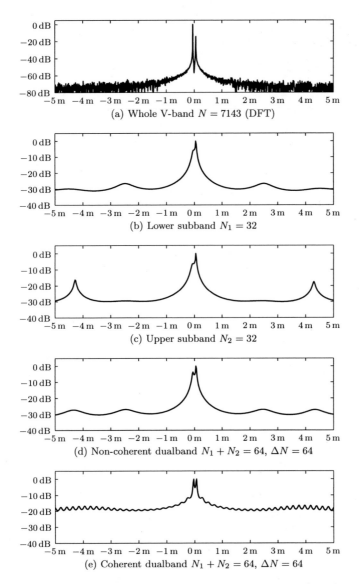

Figure 7.4: AR spectral estimation applied to the frequency response of a scenario with two cylinders (UTD simulation) at V-Band (50-75 GHz). Two subbands with bandwidth $B_l = 224\,\text{MHz}$ ($N_l = 32$) separated by a band gap of $\Delta B = 448\,\text{MHz}$ ($\Delta N = 64$) are processed with the modified covariance AR algorithm with $L = 12$.

To include the information of the two observation windows, the two approaches to build the dualband data matrix analyzed for the AR techniques are also used here for the MUSIC algorithms. It results, again, in a coherent and a non-coherent approach.

7.3.1. Non-coherent dualband MUSIC

The data matrix $\mathbf{X}_{nc}$ is defined as a vertical superposition of the subband data matrices $\mathbf{X}_{L,1}$ and $\mathbf{X}_{L,2}$

$$\mathbf{X}_{nc} = \begin{pmatrix} \mathbf{X}_{L,1} \\ \mathbf{X}_{L,2} \end{pmatrix} \tag{7.17}$$

where $\mathbf{X}_{L,1}$ and $\mathbf{X}_{L,2}$ are defined as in eq. (7.5)

$$\mathbf{X}_{L,1} = \begin{pmatrix} x[N_{10} + L] & \cdots & x[N_{10}] \\ \vdots & \ddots & \vdots \\ x[N_{10} + N_1 - 1] & \cdots & x[N_{10} + N_1 - L - 1] \end{pmatrix}$$

$$\mathbf{X}_{L,2} = \begin{pmatrix} x[N_{20} + L] & \cdots & x[N_{20}] \\ \vdots & \ddots & \vdots \\ x[N_{20} + N_2 - 1] & \cdots & x[N_{20} + N_2 - L - 1] \end{pmatrix}.$$

This data matrix is decomposed as the sum of a signal matrix and a noise matrix

$$\mathbf{X}_{nc} = \mathbf{S}_{nc} + \mathbf{W}_{nc}. \tag{7.18}$$

The signal matrix $\mathbf{S}_{nc}$ can be expressed as the product of two rank P matrices

$$\mathbf{S}_{nc} = \mathbf{B} \cdot \mathbf{C} \tag{7.19}$$

where

$$\mathbf{B} = \begin{pmatrix} h_1 z_1^{N_{10}+L} & \cdots & h_P z_P^{N_{10}+L} \\ \vdots & \ddots & \vdots \\ h_1 z_1^{N_{10}+N_1-1} & \cdots & h_P z_P^{N_{10}+N_1-1} \\ h_1 z_1^{N_{20}+L} & \cdots & h_P z_P^{N_{20}+L} \\ \vdots & \ddots & \vdots \\ h_1 z_1^{N_{20}+N_2-1} & \cdots & h_P z_P^{N_{20}+N_2-1} \end{pmatrix}$$

$$\mathbf{C} = \begin{pmatrix} 1 & z_1^{-1} & \cdots & z_1^{-L} \\ \vdots & \vdots & \ddots & \vdots \\ 1 & z_P^{-1} & \cdots & z_P^{-L} \end{pmatrix}.$$

Applying now the same analysis as in section 3.3, an eigenanalysis is done to the $P \times P$ matrix $\mathbf{B}^H \mathbf{BCC}^H$, which is positive definite and has P positive eigenvalues

$$(\mathbf{B}^H \mathbf{BCC}^H) \mathbf{v}'_k = \lambda^s_k \mathbf{v}'_k. \tag{7.20}$$

Premultiplying with the matrix $\mathbf{C}^H$ yields to

$$\mathbf{C}^H \mathbf{B}^H \mathbf{BC}(\mathbf{C}^H \mathbf{v}'_k) = \lambda^s_k (\mathbf{C}^H \mathbf{v}'_k)$$
$$\mathbf{S}^H_{nc} \mathbf{S}_{nc} \mathbf{v}_k = \lambda^s_k \mathbf{v}_k \tag{7.21}$$

where it is seen that the P nonzero eigenvalues of the matrix $\mathbf{S}^H_{nc} \mathbf{S}_{nc}$ are the eigenvalues of $\mathbf{B}^H \mathbf{BCC}^H$. The remaining eigenvalues are zero because the rank of the matrix $\mathbf{S}^H_{nc} \mathbf{S}_{nc}$ is P. The principal eigenvectors $\mathbf{v}_k$ are given by $\mathbf{C}^H \mathbf{v}'_k$, i.e. they are linear combinations of the columns of $\mathbf{C}^H$.

Considering now the noise contained in the received signal, the matrix $\mathbf{X}^H_{nc} \mathbf{X}_{nc}$ is approximated for a high number of signal samples N by its value at the limit

$$\mathbf{X}^H_{nc} \mathbf{X}_{nc} \approx \lim_{N \to \infty} \mathbf{X}^H_{nc} \mathbf{X}_{nc} = \mathbf{S}^H_{nc} \mathbf{S}_{nc} + \sigma^2_w \mathbf{I}_{L+1}. \tag{7.22}$$

Substituting $\mathbf{S}^H_{nc} \mathbf{S}_{nc}$ and $\mathbf{I}_{L+1}$ by their eigenvalue decompositions

$$\mathbf{X}^H_{nc} \mathbf{X}_{nc} = \sum_{k=1}^{P} \lambda^s_k \mathbf{v}_k + \sum_{k=1}^{L+1} \sigma^2_w \mathbf{v}_k$$
$$= \sum_{k=1}^{P} (\lambda^s_k + \sigma^2_w) \mathbf{v}_k + \sum_{k=P+1}^{L+1} \sigma^2_w \mathbf{v}_k \tag{7.23}$$

the eigenvalue decomposition of $\mathbf{X}^H_{nc} \mathbf{X}_{nc}$ is achieved. The principal eigenvectors span the *signal plus noise* subspace and they are linear combinations of the $\mathbf{s}(z_i)$, the columns of $\mathbf{C}^H$. This means that $\mathbf{s}(z_i)$, as they are spanned by the signal plus noise subspace, are orthogonal to the *noise* subspace eigenvectors or any combination of them:

$$\mathbf{s}^H(z_i) \sum_{P+1}^{L+1} \psi_k \mathbf{v}_k = 0 \tag{7.24}$$

where ψ_k are arbitrary weighting factors and

$$\mathbf{s}^H(z_i) = \begin{pmatrix} 1 & z_i^{-1} & \cdots & z_i^{-L} \end{pmatrix}$$

so that

$$\sum_{P+1}^{L+1} \psi_k \left(v_k[1] + v_k[2] z_i^{-1} + \cdots + v_k[L+1] z_i^{-L} \right) = 0 \tag{7.25}$$

and z_i are roots of any combination of the *noise polynomials*

$$V_k(z) = v_k[1] + v_k[2]z^{-1} + \cdots + v_k[L+1]z^{-L} \tag{7.26}$$

for $k = P+1, \ldots, L+1$.

An analogous analysis can be done for a signal composed by undamped exponentials $z_i = e^{j\omega_i}$. Instead of the covariance data matrix, the modified covariance data matrix is used for the analysis

$$\begin{pmatrix} \mathbf{X}_{nc} \\ \mathbf{X}_{nc}^* \mathbf{J} \end{pmatrix} \tag{7.27}$$

The corresponding signal matrix can be decomposed into two rank P matrices

$$\begin{pmatrix} \mathbf{S}_{nc} \\ \mathbf{S}_{nc}^* \mathbf{J} \end{pmatrix} = \mathbf{B} \cdot \mathbf{C} \tag{7.28}$$

in which $\mathbf{B}$ and $\mathbf{C}$ are

$$\mathbf{B} = \begin{pmatrix}
h_1 e^{j\omega_1(N_{10}+L)} & \cdots & h_P e^{j\omega_P(N_{10}+L)} \\
\vdots & \ddots & \vdots \\
h_1 e^{j\omega_1(N_{10}+N_1-1)} & \cdots & h_P e^{j\omega_P(N_{10}+N_1-1)} \\
h_1 e^{j\omega_1(N_{20}+L)} & \cdots & h_P e^{j\omega_P(N_{20}+L)} \\
\vdots & \ddots & \vdots \\
h_1 e^{j\omega_1(N_{20}+N_2-1)} & \cdots & h_P e^{j\omega_P(N_{20}+N_2-1)} \\
h_1^* e^{-j\omega_1 N_{10}} & \cdots & h_P^* e^{-j\omega_P N_{10}} \\
\vdots & \ddots & \vdots \\
h_1^* e^{-j\omega_1(N_{10}+N_1-L-1)} & \cdots & h_P^* e^{-j\omega_P(N_{10}+N_1-L-1)} \\
h_1^* e^{-j\omega_1 N_{20}} & \cdots & h_P^* e^{-j\omega_P N_{20}} \\
\vdots & \ddots & \vdots \\
h_1^* e^{-j\omega_1(N_{20}+N_2-L-1)} & \cdots & h_P^* e^{-j\omega_P(N_{20}+N_2-L-1)}
\end{pmatrix}$$

$$\mathbf{C} = \begin{pmatrix}
1 & e^{-j\omega_1} & \cdots & e^{-j\omega_1 L} \\
\vdots & \vdots & \ddots & \vdots \\
1 & e^{-j\omega_P} & \cdots & e^{-j\omega_P L}
\end{pmatrix}.$$

Applying the same eigenanalysis as for the covariance data matrix in eq. (7.17), it can be stated that $\mathbf{s}(e^{j\omega_i})$, the columns of $\mathbf{C}^H$, are orthogonal to the noise subspace eigenvectors or any combination of them:

$$\mathbf{s}^H(e^{j\omega_i}) \sum_{P+1}^{L+1} \psi_k \mathbf{v}_k = 0 \tag{7.29}$$

where ψ_k are arbitrary weighting factors and

$$\mathbf{s}^H(e^{j\omega_i}) = \begin{pmatrix} 1 & e^{-j\omega_i} & \cdots & e^{-j\omega_i L} \end{pmatrix}$$

so that

$$\sum_{P+1}^{L+1} \psi_k \left(v_k[1] + v_k[2]e^{-j\omega_i} + \cdots + v_k[L+1]e^{-j\omega_i L} \right) = 0 \qquad (7.30)$$

and $e^{j\omega_i}$ are roots of any combination of the noise polynomials

$$V_k(z) = v_k[1] + v_k[2]z^{-1} + \cdots + v_k[L+1]z^{-L} \qquad (7.31)$$

for $k = P + 1, \ldots, L + 1$.

It is noted that the same noise polynomial structure as in the single-band case has been obtained with this non-coherent approach. The gap information is not used, and therefore, non-coherent data sets can be used. Due to the structure of the data matrix, the performance of the estimator is expected to be the same as for the single-band algorithm with $N_1 + N_2$ data samples. Therefore, this non-coherent approach results in a robust way to use non-coherent data sets for spectrum estimation and achieves the same performance as the single-window algorithm with an equivalent data set length equal to the sum of the subwindows lengths. Also, different amplitudes at different observation windows does not become a problem, as it can be seen from the structures of the matrices $\mathbf{B}$ and $\mathbf{C}$. As drawback, no advantage of the gap information can be obtained.

7.3.2. Coherent dualband MUSIC

The data matrix $\mathbf{X}_c$ is built as the horizontal juxtaposition of the subband matrices $\mathbf{X}_{L_1,1}$ and $\mathbf{X}_{L_2,2}$

$$\mathbf{X}_c = \begin{pmatrix} \mathbf{X}_{L_2,2} & \mathbf{X}_{L_1,1} \end{pmatrix} \qquad (7.32)$$

where

$$\mathbf{X}_{L_1,1} = \begin{pmatrix} x[N_{10} + L_1] & \cdots & x[N_{10}] \\ \vdots & \ddots & \vdots \\ x[N_{10} + N_1 - 1] & \cdots & x[N_{10} + N_1 - L_1 - 1] \end{pmatrix}$$

$$\mathbf{X}_{L_2,2} = \begin{pmatrix} x[N_{20} + L_2] & \cdots & x[N_{20}] \\ \vdots & \ddots & \vdots \\ x[N_{20} + N_2 - 1] & \cdots & x[N_{20} + N_2 - L_2 - 1] \end{pmatrix}$$

and the following relation has been assumed

$$N_2 - N_1 = L_2 - L_1. \qquad (7.33)$$

The data matrix is decomposed as the sum of a signal matrix and a noise matrix

$$\mathbf{X}_c = \mathbf{S}_c + \mathbf{W}_c. \tag{7.34}$$

The signal matrix $\mathbf{S}_c$ can be expressed as the product of two rank P matrices

$$\mathbf{S}_c = \mathbf{B} \cdot \mathbf{C} \tag{7.35}$$

where

$$\mathbf{B} = \begin{pmatrix} h_1 z_1^{N_{20}+L_2} & \cdots & h_P z_P^{N_{20}+L_2} \\ \vdots & \ddots & \vdots \\ h_1 z_1^{N_{20}+N_2-1} & \cdots & h_P z_P^{N_{20}+N_2-1} \end{pmatrix}$$

$$\mathbf{C} = \begin{pmatrix} 1 & z_1^{-1} & \cdots & z_1^{-L_2} & z_1^{-M} & \cdots & z_1^{-M-L_1} \\ \vdots & \vdots & \ddots & \vdots & \vdots & \ddots & \vdots \\ 1 & z_P^{-1} & \cdots & z_P^{-L_2} & z_P^{-M} & \cdots & z_P^{-M-L_1} \end{pmatrix}$$

with

$$M = N_{20} - N_{10} + N_2 - N_1.$$

Following the same analysis as in the non-coherent approach, the eigenvalue decomposition of the matrix $\mathbf{X}_c^H \mathbf{X}_c$ is obtained

$$\begin{aligned} \mathbf{X}_c^H \mathbf{X}_c &= \sum_{k=1}^{P} \lambda_k^s \mathbf{v}_k + \sum_{k=1}^{L+2} \sigma_w^2 \mathbf{v}_k \\ &= \sum_{k=1}^{P} (\lambda_k^s + \sigma_w^2) \mathbf{v}_k + \sum_{k=P+1}^{L+2} \sigma_w^2 \mathbf{v}_k \end{aligned} \tag{7.36}$$

where $L = L_1 + L_2$. The principal P eigenvectors span the signal plus noise subspace and they are a linear combination of $\mathbf{s}(z_i)$, the columns of $\mathbf{C}^H$. Therefore, the remaining $L+2-P$ eigenvectors—which span the noise subspace—or any combination of them will be orthogonal to $\mathbf{s}(z_i)$

$$\mathbf{s}(z_i)^H \cdot \sum_{k=P+1}^{L+2} \psi_k \mathbf{v}_k = 0 \tag{7.37}$$

where ψ_k are arbitrary weighting factors and

$$\mathbf{s}^H(z_i) = \begin{pmatrix} 1 & z_i^{-1} & \cdots & z_i^{-L_2} & z_i^{-M} & \cdots & z_i^{-M-L_1} \end{pmatrix}.$$

Rewriting the relation into polynomial form

$$\sum_{P+1}^{L+2} \psi_k \left(v_k[1] + v_k[2]z_i^{-1} + \cdots + v_k[L_2 + 1]z_i^{-L_2} + \right.$$
$$\left. + v_k[L_2 + 2]z_i^{-M} + \cdots + v_k[L + 2]z_i^{-M-L_1} \right) = 0 \quad (7.38)$$

it is noted that z_i are roots of any combination of the noise polynomials

$$V_k(z) = v_k[1] + v_k[2]z^{-1} + \cdots + v_k[L_2 + 1]z^{-L_2} +$$
$$+ v_k[L_2 + 2]z^{-M} + \cdots + v_k[L + 2]z^{-M-L_1} \quad (7.39)$$

for $k = P + 1, \ldots, L + 2$.

An analogous analysis can be done for a signal composed by undamped exponentials $z_i = e^{j\omega_i}$. Instead of the covariance data matrix, the modified covariance data matrix is used for the analysis

$$\begin{pmatrix} \mathbf{X}_c \\ \mathbf{X}_c^* \mathbf{J} \end{pmatrix} \quad (7.40)$$

The corresponding data matrix is decomposed into two rank P matrices

$$\begin{pmatrix} \mathbf{S}_c \\ \mathbf{S}_c^* \mathbf{J} \end{pmatrix} = \mathbf{B} \cdot \mathbf{C} \quad (7.41)$$

where

$$\mathbf{B} = \begin{pmatrix} h_1 e^{j\omega_1(N_{20}+L_2)} & \cdots & h_P e^{j\omega_P(N_{20}+L_2)} \\ \vdots & \ddots & \vdots \\ h_1 e^{j\omega_1(N_{20}+N_2-1)} & \cdots & h_P e^{j\omega_P(N_{20}+N_2-1)} \\ h_1^* e^{-j\omega_1 N_{10}} & \cdots & h_P^* e^{-j\omega_P N_{10}} \\ \vdots & \ddots & \vdots \\ h_1^* e^{-j\omega_1(N_{10}+N_1-L-1)} & \cdots & h_P^* e^{-j\omega_P(N_{10}+N_1-L-1)} \end{pmatrix}$$

$$\mathbf{C} = \begin{pmatrix} 1 & e^{-j\omega_1} & \cdots & e^{-j\omega_1 L_2} & e^{-j\omega_1 M} & \cdots & e^{-j\omega_1(M+L_1)} \\ \vdots & \vdots & \ddots & \vdots & \vdots & \ddots & \vdots \\ 1 & e^{-j\omega_P} & \cdots & e^{-j\omega_P L_2} & e^{-j\omega_P M} & \cdots & e^{-j\omega_P(M+L_1)} \end{pmatrix}.$$

Applying the same eigenanalysis as for the covariance data matrix in eq. (7.32), it can be stated that $\mathbf{s}(e^{j\omega_i})$, the columns of $\mathbf{C}^H$, are orthogonal to the noise subspace eigenvectors or any combination of them

$$\mathbf{s}^H(e^{j\omega_i}) \sum_{P+1}^{L+2} \psi_k \mathbf{v}_k = 0 \quad (7.42)$$

where ψ_k are arbitrary weighting factors and

$$\mathbf{s}^H(e^{j\omega_i}) = \begin{pmatrix} 1 & e^{-j\omega_i} & \cdots & e^{-j\omega_i L_2} & e^{-j\omega_i M} & \cdots & e^{-j\omega_i(M+L_1)} \end{pmatrix}.$$

Rewriting the relation into polynomial form

$$\sum_{P+1}^{L+2} \psi_k \left(v_k[1] + v_k[2]e^{-j\omega_i} + \cdots + v_k[L_2+1]e^{-j\omega_i L_2} + \right.$$
$$\left. + v_k[L_2+2]e^{-j\omega_i M} + \cdots + v_k[L+2]e^{-j\omega_i(M+L_1)} \right) = 0 \quad (7.43)$$

it is noted that $e^{-j\omega_i}$ are roots of any combination of the noise polynomials

$$V_k(z) = v_k[1] + v_k[2]z^{-1} + \cdots + v_k[L_2+1]z^{-L_2} +$$
$$+ v_k[L_2+2]z^{-M} + \cdots + v_k[L+2]z^{-M-L_1} \quad (7.44)$$

for $k = P+1, \ldots, L+2$.

It is noted that the gap information is now involved in the polynomials. Of course, as the gap information is required, it must be known exactly, i.e. coherence between the data must be assured. The main drawback of the method is the linear relation between the order of the noise polynomial and the gap. The order is now $M+L_1$ and this results in the appearance of a higher number of zeros which difficult the selection of the ones corresponding to the signal. This may be solved by combining the coherent with the non-coherent approaches, as it will be shown in section 7.4.

7.3.3. Spectral MUSIC

Based on the eigenanalysis decomposition, the spectral MUSIC, looks for the zeros of the noise polynomials by evaluating the peaks of the pseudo-spectrum:

$$P_{\text{MUSIC}}(f) = \frac{1}{D(e^{j2\pi f})} = \frac{1}{\mathbf{s}^H(e^{j\omega}) \left(\sum_{k>P} \mathbf{v}_k \mathbf{v}_k^H \right) \mathbf{s}(e^{j\omega})} \quad (7.45)$$

where the structure of the frequency vector $\mathbf{s}(e^{j\omega})$ depends on the selected approach: for the non-coherent algorithm

$$\mathbf{s}(e^{j\omega}) = \begin{pmatrix} 1 & e^{j\omega} & \cdots & e^{j\omega L} \end{pmatrix}^T \quad (7.46)$$

and for the coherent algorithm

$$\mathbf{s}(e^{j\omega}) = \begin{pmatrix} 1 & e^{j\omega} & \cdots & e^{j\omega L_2} & e^{j\omega M} & \cdots & e^{j\omega(M+L_1)} \end{pmatrix}^T. \quad (7.47)$$

7.3.4. Root MUSIC

The roots of the null spectra polynomial are directly obtained:

$$D(z) = \sum_{k>P} V_k(z)V_k^*(1/z^*) \tag{7.48}$$

where the noise polynomials depend on the selected approach: for the non-coherent algorithm

$$V_k(z) = v_k[1] + v_k[2]z^{-1} \cdots + v_k[L+1]z^{-L} \tag{7.49}$$

and for the coherent approach

$$V_k(z) = v_k[1] + \cdots + v_k[L_2+1]z^{-L_2} + \\ v_k[L_2+2]z^{-M} + \cdots + v_k[L+2]z^{-M-L_1}. \tag{7.50}$$

The roots z_j will appear together with their reflected $1/z_j^*$ and only the frequencies determined by the roots on or inside the unit circle are taken into account.

7.3.5. Conclusions

Two main approaches to apply MUSIC spectral analysis to dualband signals have been presented: coherent and non-coherent, where the information of the band gap is required or not, respectively.

The non-coherent approach is a robust way to use multiband data sets and achieves the same performance as the single-band algorithm with an equivalent data set length equal to the sum of the subbands lengths. As drawback, no advantage of the gap information can be obtained.

In the coherent approach, the relation between data samples of different subbands is explicitly used and appears in the data matrix structure. As in the AR case, the main drawback is the linear increase of possible targets with increasing band gap, which complicates the selection of the targets corresponding to the received signal. Combining the coherent with the non-coherent approaches, as shown in section 7.4, can be used to solve this problem.

Figure 7.5 shows different MUSIC range profile estimates applied to the frequency response of a two-cylinder scenario—as in figure 7.4 for the AR analysis—obtained with the UTD approximation as described in appendix A. The single-band and the non-coherent dualband estimates do not resolve the two cylinders. The presence of two targets can be guess, but they are not clearly resolved. In the coherent dualband estimate the two cylinders can be identified. Also a ripple due to the high-order polynomial involved in the algorithm can be observed. As in the AR case, this undesired ripple disturbs the estimate and can origin ambiguity problems.

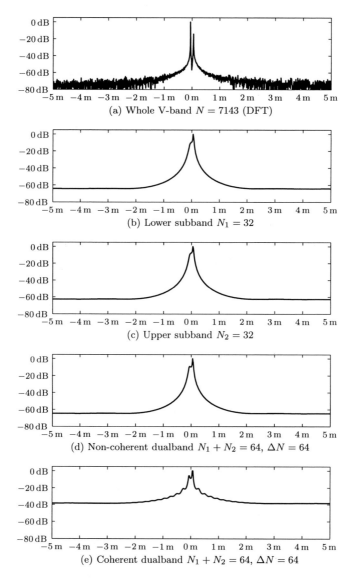

Figure 7.5: MUSIC spectral estimation applied to the frequency response of a scenario with two cylinders (UTD simulation) at V-Band (50-75 GHz). Two subbands with bandwidth $B_l = 224$ MHz ($N_l = 32$) separated by a band gap of $\Delta B = 448$ MHz ($\Delta N = 64$) are processed with the MUSIC algorithm using the modified covariance data matrix and parameters $L = 12$ and $P = 2$.

7.4. Combination of the dualband coherent and non-coherent algorithms

The presented dualband non-coherent algorithms perform a robust estimation of the poles contained in the signal. This algorithms process data sets without analyzing the band gap information, so that coherence between the data sets is not required. However, their performance is expected to be the same as an equivalent single-band algorithm and not to improve with increasing band gap. On the other hand, the coherent algorithms use the band gap information and a benefit from this additional information is expected. However, an ambiguity problem appears because of the high order of the polynomials involved in the estimation. The main capabilities of this two approaches can be combined to obtain a robust estimation of the signal poles with increased performance.

From the whole set of poles obtained with the coherent algorithms only those in the vicinity of the signal poles obtained with the non-coherent algorithms are analyzed. The search radius should be lower than the unambiguous range, which is given by the order of the polynomial in the coherent approaches. The polynomial to solve has a total number of zeros of approx. $N_{20} - N_{10}$, assuming that they are distributed uniformly along the unit circle, the distance between two poles will be

$$f_{\text{unamb}} = \frac{1}{N_{20} - N_{10}} \tag{7.51}$$

This search radius gives the limit of this combination of information, as this radius decreases with increasing band gap.

8. Frequency band gap estimation

In previous chapters was shown, that the use of the band gap information in a dualband radar sensor system can improve the range accuracy. For closely spaced targets—closer than the Rayleigh distance—it improves even the range resolution in some cases. To use the band gap information a coherent data set is required, i. e. the relative distance between observation windows must be known exactly. The coherence between the two sensors, however, implies a higher cost for the system. In this chapter, a way to estimate the band gap between two non-coherent sensors is presented.

8.1. Method

Assuming a radar signal as presented in chapter 2, a radar scenario observed in the frequency domain can be modeled by the sum of complex exponentials:

$$s[n] = \sum_i h_i z_i^n \tag{8.1}$$

where $z_i = \rho_i e^{j2\pi f_i}$ represents a scattering center at the normalized distance f_i, and with a frequency amplitude dependence determined by ρ_i. The scene is observed at two frequency bands $\mathcal{B}_l$ under the same aspect angle, so that the independent variable n is expressed as

$$n = \begin{cases} N_{10}, \ldots, N_{10} + N_1 - 1 & \text{at } \mathcal{B}_1 \\ N_{20}, \ldots, N_{20} + N_2 - 1 & \text{at } \mathcal{B}_2 \end{cases} \tag{8.2}$$

and the poles are the same for both observation windows.

The radar scenario is measured with two independent non-coherent systems, so that measuring errors in amplitude and distance can appear. This type of errors can be corrected by proper calibration techniques, such as the technique presented in appendix B. Coherence errors, however, remain after calibration. An approximate knowledge of the distance $\Delta N' = N_{20} - N_{10}$ is assumed, so that to the known distance $\Delta N'$ an unknown error ϵ_Δ is added. The signal is evaluated at the sample points:

$$n = \begin{cases} N_{10}, \ldots, N_{10} + N_1 - 1 & \text{at } \mathcal{B}_1 \\ N_{20} + \epsilon_\Delta, \ldots, N_{20} + \epsilon_\Delta + N_2 - 1 & \text{at } \mathcal{B}_2 \end{cases} \tag{8.3}$$

and the measured radar signal has the expression

$$s[n] = \begin{cases} \sum_i h_i z_i^{N_{10}} z_i^{k_1} & \text{at } \mathcal{B}_1 \\ \sum_i h_i z_i^{N_{20}+\epsilon_\Delta} z_i^{k_2} & \text{at } \mathcal{B}_2 \end{cases} \tag{8.4}$$

where

$$k_1 = 0, \ldots, N_1 - 1$$
$$k_2 = 0, \ldots, N_2 - 1$$

This expression can be written as

$$s[n] = \begin{cases} \sum_i h_{i,1} z_i^{k_1} & \text{at } \mathcal{B}_1 \\ \sum_i h_{i,2} z_i^{k_2} & \text{at } \mathcal{B}_2 \end{cases} \tag{8.5}$$

and the relation between both complex amplitudes results in

$$\frac{h_{i,2}}{h_{i,1}} = z_i^{N_{20}+\epsilon_\Delta-N_{10}} = z_i^{\Delta N'+\epsilon_\Delta}. \tag{8.6}$$

It is noted that the unknown ϵ_Δ is included in the phase term of this quotient together with the known $\Delta N'$. To isolate the unknown parameter, the following expression is evaluated

$$\arg\left\{ \frac{h_{i,2}}{h_{i,1}} z_i^{-\Delta N'} \right\} = 2\pi f_i \epsilon_\Delta. \tag{8.7}$$

For coherent data sets, the phase term of an oscillation in one observation window continues without *phase jump* in the next one, so that the previous expression would be zero, which is consistent with $\epsilon_\Delta = 0$. For non-coherent data sets, however, the phase jump will depend on the window shift ϵ_Δ and on the normalized distance f_i. This relation can be used to estimate ϵ_Δ. Of course, the poles are in principle also unknown, but they can be estimated using the dualband non-coherent techniques presented in chapter 7, which are not affected by coherence errors. Once the poles are obtained, the optimum amplitudes—in a minimum square error sense—for each window can be obtained solving a linear system of equations, so that the quotient in eq. (8.7) can be calculated. For each pole in the signal there is a relation for the window shift, i. e. the system of equations is overdetermined. Due to noise and limited data set lengths, some errors appear in the estimation of the poles so that linear regression can be used to find the ϵ_Δ which best fits the different amplitude relations, as shown in figure 8.1.

Each pole in the signal is assumed to have different energy, so that the error in its estimate also varies. Assuming lower estimation error for poles with greater energy, the pole energy can be used as weighting factor in the linear regression to estimate ϵ_Δ. This permits to apply this approach to realistic data, so that targets with lower energy influence less the ϵ_Δ estimation.

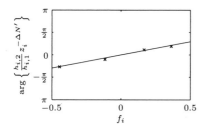

Figure 8.1: Illustration of the relation between ϵ_Δ and the quotient in eq. (8.7). A linear regression is used to select the ϵ_Δ which best fits to the set of quotients

As the relation in eq. (8.7) is based on a phase function, an ambiguity in the estimation of ϵ_Δ appears. Assuming a limited domain for the normalized distance $-0.5 < f_i < 0.5$, and a limited range of the argument function $-\pi < \arg\{x\} < \pi$, the maximum band gap deviation which can be estimated unambiguously must be lower than one sample $-1 < \epsilon_\Delta < 1$.

For coherent data sets the relation in eq. (8.7) can be used to obtain a better estimate of the poles by modifying slightly f_i to impose a zero phase in the quotient.

8.2. Monte Carlo analysis

To evaluate the performance of this approach a Monte Carlo analysis with 5000 trials is realized. A signal with P poles is sampled at two observation windows with equal length $N_1 = N_2 = 32$. The poles lie on the unit circle and the normalized distances are uniformly distributed random variables $f_i \sim \mathcal{U}(-0.5, 0.5)$. Just a minimum distance between the targets of two times the Rayleigh limit, i.e. $2/(N_1 + N_2)$, is assured. Each pole has an independent random complex Gaussian amplitude $h_i \sim \mathcal{CN}(0, \sigma^2)$. The signal is observed in the presence of *complex white Gaussian noise* (CWGN) $w[n] \sim \mathcal{CN}(0, \sigma_w^2)$ and the signal-to-noise ratio is defined as SNR $= \sigma^2/\sigma_w^2$.

To obtain the signal poles a non-coherent modified covariance AR algorithm with $L = 12$ is applied. The overdetermined system of equations for ϵ_Δ given in eq. (8.7) is solved by minimizing the following error

$$\sum_i \left| \left(\arg\left\{ \frac{h_{i,2}}{h_{i,1}} e^{-j2\pi f_i \Delta N'} \right\} - 2\pi f_i \epsilon_\Delta \right) \cdot h_{i,1} \cdot h_{i,2} \right|^2 . \tag{8.8}$$

It is seen that the amplitude of the poles are used as weighting factors, and that the pole estimates are forced to lie on the unit circle. The pole amplitudes $h_{i,l}$ are obtained with the *least square error* (LSE) criteria.

Results of the simulations are shown in figures 8.2 and 8.3 for $P = 4$ and $P = 6$ poles. Figure 8.2 shows the mean and the variance as a function of the gap ΔN, and figure 8.3 as function of the signal to noise ratio SNR. It can be

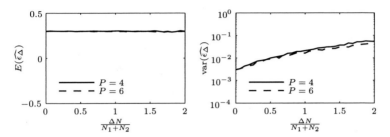

Figure 8.2: Mean value and variance of the ϵ_Δ estimate as function of the gap ΔN for SNR = 10 dB and $\epsilon_\Delta = 0.3$

observed that the estimate can be considered unbiased, only a small deviation appear for SNR lower than 10 dB. The variance is slightly lower for $P = 6$ than for $P = 4$, it increases with the gap ΔN and, as expected, decreases with increasing SNR.

If the variance of the estimate is not low enough for a certain application, an improvement can be achieved by averaging several consecutive estimates, as the band gap should remain constant. Also adaptive approach as used in many other control or signal processing applications (see e.g. [29, 95]) can be used to follow possible variations of ϵ_Δ with time.

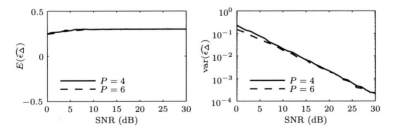

Figure 8.3: Mean value and variance of the ϵ_Δ estimate as function of SNR for $\Delta N = 64$ and $\epsilon_\Delta = 0.3$

9. Monte Carlo analysis of multiband radar signal processing

9.1. Description

To assess the range resolution and accuracy performance which can be obtained with the multiband spectral estimation algorithms presented in chapter 7, Monte Carlo analyses with 5000 trials have been carried out for different signal scenarios. To assure controlled parameters in all the experiments synthetic data has been used. It provides also greater flexibility than measured or simulated data sets in the selection of the scenarios, and allows the randomization of some variables.

A radar scenario observed in the frequency domain is considered for the analysis. The scenario is decomposed, as described in chapter 2, as a sum of complex oscillations, each representing a target at normalized distance f_i with complex amplitude h_i

$$s[n] = \sum_i h_i e^{j2\pi f_i}. \tag{9.1}$$

Signal samples are available at two different observation windows $\mathcal{B}_l$, i. e. frequency bands, so that the independent variable is expressed as

$$n = \begin{cases} N_{10}, \ldots, N_{10} + N_1 - 1 & \text{at } \mathcal{B}_1 \\ N_{20}, \ldots, N_{20} + N_2 - 1 & \text{at } \mathcal{B}_2 \end{cases} \tag{9.2}$$

Multiband coherent and non-coherent AR and Root-MUSIC algorithms based on the covariance data matrix are used for the analysis. The obtained results, however, can be extended to other approaches. Results with the single-band versions of these algorithms are also included in the analysis, assuming a total number of observed samples equal to the sum of the samples in the subwindows. The number of oscillations present in the signal is assumed to be known. This is in practice not real. Well-known algorithms to estimate the number of sinusoids can be applied [2, 3, 21, 62, 67, 93].

To analyze the resolution capability of any spectral estimation algorithm, it is necessary to define the resolution event, i. e. a criteria to decide, based on the estimator output, if two targets are resolved or not. This criteria must be suitable for the type of estimator, so that for this analysis the decision must be based on a set of poles. The definition used is illustrated in figure 9.1:

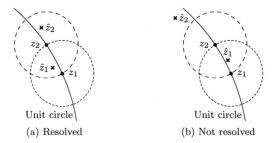

(a) Resolved   (b) Not resolved

Figure 9.1: Illustration of the resolution event

two poles z_1 and z_2 are considered to be *resolved* if the absolute error for both estimates, $\hat{z}_1$ and $\hat{z}_2$, is lower than the distance between the true poles $|z_1 - z_2|$

$$
\begin{aligned}
\text{resolved} \quad &: \quad |z_1 - \hat{z}_1| < |z_1 - z_2| \wedge |z_2 - \hat{z}_2| < |z_1 - z_2| \\
\text{not resolved} \quad &: \quad |z_1 - \hat{z}_1| > |z_1 - z_2| \vee |z_2 - \hat{z}_2| > |z_1 - z_2|
\end{aligned}
\tag{9.3}
$$

The poles which are nearer to the real poles z_i are selected as estimates $\hat{z}_i$. This definition for the resolution event suits very well for algorithms with a set of poles as output. With this criteria not only the angle information but also the distance to the unit circle is taken into account. Poles with an absolute value too far away from the unit circle, i.e. corresponding to less sharp peaks in the spectrum, will hardly fulfill the resolution criteria. It can also be noted that for wider spaced poles, the permitted estimation error increases. This makes also sense as the accuracy required to distinguish between two poles must be in general lower than their distance.

Resolution and accuracy capabilities are obtained in different two-target scenarios. An analysis of the *target split probability* (P_{split}), i.e. the probability of detecting two targets when only one target is present in the signal, and of the sensitivity to coherence errors complete the assessment.

For the resolution assessment, scenarios with two scatterers with equal energy at normalized distances f_i are considered. Without loss of generality, the absolute value of the amplitudes is set to one $|h_i| = 1$

$$
s[n] = e^{j\phi_1} e^{j2\pi f_1} + e^{j\phi_2} e^{j2\pi f_2}.
\tag{9.4}
$$

The amplitude phase is considered as a random variable with a uniform distribution $\phi_i \sim \mathcal{U}(0, 2\pi)$. The frequency distance Δf is constant and controlled in the simulations, but not the center frequency $f_c = (f_1 + f_2)/2$, which is also a random variable with an uniform distribution $f_c \sim \mathcal{U}(-0.5, 0.5)$. This radar signal is received in the presence of additive complex white Gaussian noise (CWGN), $w[n] \sim \mathcal{CN}(0, \sigma_w^2)$, so that

$$
x[n] = s[n] + w[n]
\tag{9.5}
$$

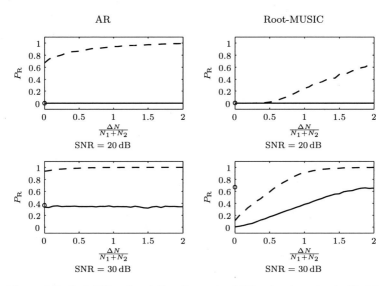

Figure 9.2: Probability of resolution P_R vs. gap ΔN in a two-tone signal with $\Delta f = \frac{0.2}{N_1+N_2}$ for single-band (circle), dualband non-coherent (solid line) and dualband coherent (dashed line) AR and Root-MUSIC algorithms

and the signal-to-noise ratio (SNR) is defined as SNR $= 1/\sigma_w^2$.

The observation subwindows have equal length $N_1 = N_2 = 32$ and variable window gap $\Delta N = N_{20} - N_{10} - N_1 = \Delta N' - N_1$. The selected parameters for the spectral estimators are $L = 10$ for the AR algorithm, and $L = 10$ and $P = 2$ for the Root-MUSIC algorithm.

For the P_{split} analysis, one-target scenarios are considered. The absolute value of the target amplitude is set again to one $|h_c| = 1$

$$s[n] = e^{j\phi_c}e^{j2\pi f_c} \tag{9.6}$$

and the phase is a uniformly distributed random variable $\phi_c \sim \mathcal{U}(0, 2\pi)$. The frequency is also a random variable with uniform distribution $f_c \sim \mathcal{U}(-0.5, 0.5)$. These one-target scenarios are analyzed in the same way as for the resolution analysis: the AR and MUSIC algorithms are applied with the same parameters, i.e. $L = 10$ and $P = 2$, and the resolution criteria is evaluated expecting two oscillations at $f_c - \Delta f/2$ and $f_c + \Delta f/2$. The cases where the criteria decides that two targets have been resolved, are considered target splits.

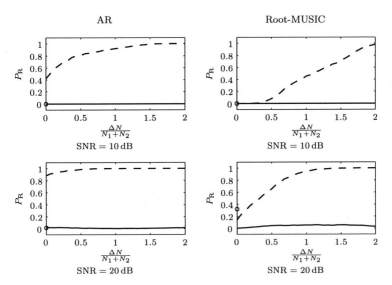

Figure 9.3: Probability of resolution P_R vs. gap ΔN in a two-tone signal with $\Delta f = \frac{0.3}{N_1+N_2}$ for single-band (circle), dualband non-coherent (solid line) and dualband coherent (dashed line) AR and Root-MUSIC algorithms

9.2. Results

In figures 9.2 to 9.6 the probability of resolution P_R is displayed as a function of the window gap ΔN for two-target scenarios with different Δf and SNR. To obtain correct conclusions about P_R, it is necessary to analyze also the target split probability P_split for one-target scenarios, shown in figure 9.7, in order to distinguish between resolution effects and ambiguous detections.

Figure 9.2 shows the resolution performance for scenarios with two closely spaced tones, i.e. $\Delta f = 0.2/(N_1 + N_2)$. It is observed the better resolution performance of the coherent algorithms, and the P_R improvement with increasing window gap ΔN. The dependence on the window gap is greater in the Root-MUSIC as in the AR algorithm, which shows also a better performance. It is also noted that the performances of the single-band and the dualband non-coherent AR algorithms are equal, which is expected from the structure of the data matrix used in the approaches. In the Root-MUSIC algorithms, however, the single-band approach presents a better performance as the dualband non-coherent one.

In figure 9.3 the performance for scenarios with slightly wider spaced targets, i.e. $\Delta f = 0.3/(N_1 + N_2)$, is observed. The comments done for the previous case are also valid here. Analyzing figure 9.7 for $\Delta f = 0.3/(N_1 + N_2)$, it is

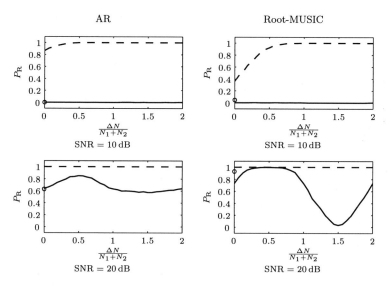

Figure 9.4: Probability of resolution P_R vs. gap ΔN in a two-tone signal with $\Delta f = \frac{0.5}{N_1+N_2}$ for single-band (circle), dualband non-coherent (solid line) and dualband coherent (dashed line) AR and Root-MUSIC algorithms

seen that for window gaps greater than $1.5 \cdot (N_1 + N_2)$ ambiguity problems arise with the coherent dualband approaches. An analysis of the resolution beyond this bound is, therefore, difficult.

The resolution performance and target split behavior for two-target scenarios with $\Delta f = 0.5/(N_1 + N_2)$ are shown in figures 9.4 and 9.7, respectively. Target splitting arises for window gaps greater than $0.6 \cdot (N_1 + N_2)$ in the coherent approach, so that an analysis of the P_R makes only sense for lower window gaps. As in the previous cases, the Root-MUSIC algorithms show a greater dependence on the gap.

Figure 9.5 displays the performance of the algorithms in scenarios with $\Delta f = 1/(N_1 + N_2)$. As seen in figure 9.7, ambiguity problems in the coherent approaches arise for very low window gaps, so that an analysis of the resolution performance is only meaningful for the non-coherent algorithms. A periodic behavior of P_R with period $1/\Delta f$ is observed. As in previous scenarios, a greater dependence on the window gap is observed by the Root-MUSIC algorithm, whereas the performance in all approaches is very similar.

For the analysis in scenarios with target separation two times the Rayleigh bound, i. e. $\Delta f = 2/(N_1 + N_2)$, no dualband coherent analysis has been done. As in the scenarios with $\Delta f = 1/(N_1 + N_2)$, only ambiguous targets would be detected with the dualband coherent approaches. Figure 9.6 displays the

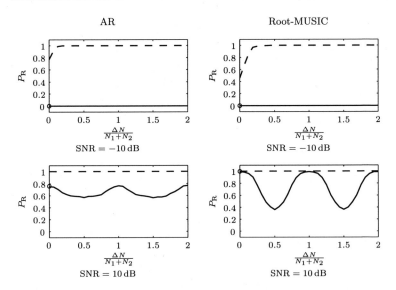

Figure 9.5: Probability of resolution P_R vs. gap ΔN in a two-tone signal with $\Delta f = \frac{1}{N_1+N_2}$ for single-band (circle), dualband non-coherent (solid line) and dualband coherent (dashed line) AR and Root-MUSIC algorithms

performance of the single-band and dualband non-coherent algorithms. Again, a slight better performance is obtained with the single-band algorithms. The oscillating behavior with period $1/\Delta f$ is still observed in P_R, although the dependence is lower than in scenarios with lower target separation.

The oscillating behavior of P_R with period $1/\Delta f$ observed in the scenarios was already predicted in the theoretical resolution performance bounds obtained with the hypothesis testing theory. It can be concluded that the window gap does not improve continuously the resolution performance. An oscillating behavior is obtained with period depending on the target separation, i.e. $1/\Delta f$. For greater distances, however, the performance does almost not change with the window gap.

Figure 9.8 shows the variance of the estimates for different Δf. In all cases the SNR is high enough to ensure resolution, otherwise, the analysis of the variance would not be really meaningful. The dependence of $\text{var}(\hat{f}_i)$ on the window gap has also a periodic pattern with distance between notches or peaks equal to $1/\Delta f$. As in the resolution performance, the dependence on ΔN is lower for scenarios with greater Δf.

An analysis of the sensitivity of the algorithms to coherence errors completes the analysis of the dualband spectral estimators. The coherence error

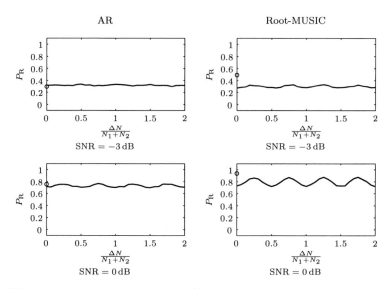

Figure 9.6: Probability of resolution P_R vs. gap ΔN in a two-tone signal with $\Delta f = \frac{2}{N_1+N_2}$ for single-band (circle) and dualband non-coherent (solid line) AR and Root-MUSIC algorithms

is expressed as a shift in the observation window, so that the signal is actually evaluated at the index

$$n = \begin{cases} N_{10}, \ldots, N_{10} + N_1 - 1 & \text{at } \mathcal{B}_1 \\ N_{20} + \epsilon_\Delta, \ldots, N_{20} + \epsilon_\Delta + N_2 - 1 & \text{at } \mathcal{B}_2 \end{cases} \qquad (9.7)$$

Figures 9.9 and 9.10 show the resolution and the accuracy performance, respectively. As expected, the performance of the non-coherent algorithms is not affected by ϵ_Δ. On the contrary, the performance of the coherent approaches is influenced by the error ϵ_Δ.

9.3. Conclusions

In this chapter a Monte Carlo analysis for the resolution and accuracy performance of dualband spectral estimators has been presented. The behavior predicted with the hypothesis testing theory, i.e. a periodic pattern for P_R and $\mathrm{var}(\hat{f}_i)$ with period $1/\Delta f$ has been confirmed. The window gap does not improve continuously the resolution performance, an oscillating behavior with period $1/\Delta f$ is obtained. For greater distances, however, the performance does

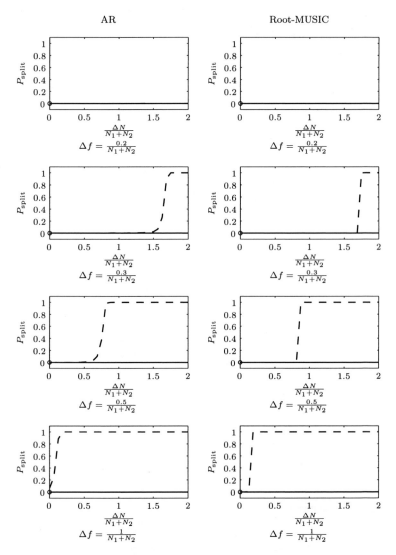

Figure 9.7: Target split probability P_{split} vs. gap ΔN in a single-tone signal with SNR = 30 dB for single-band (circle), dualband non-coherent (solid line) and dualband coherent (dashed line) AR and Root-MUSIC algorithms. Resolution criteria applied with different Δf

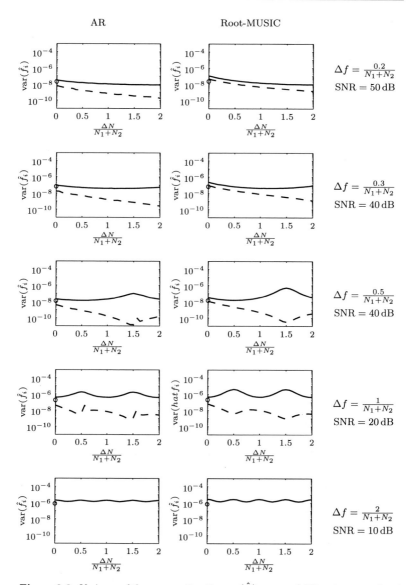

Figure 9.8: Variance of the range estimation var($\hat{f}_i$) vs. gap ΔN in a two-tone signal with single-band (circle), dualband non-coherent (solid line) and dualband coherent (dashed line) AR and Root-MUSIC algorithms

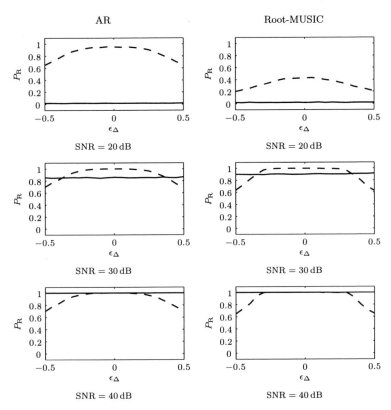

Figure 9.9: Probability of resolution P_R as a function of the gap error ϵ_Δ in two-tone scenarios with $\Delta f = \frac{0.3}{N_1+N_2}$, sampled in a dualband system with $\Delta N = 0.25 \cdot (N_1 + N_2)$ for dualband non-coherent (solid line) and dualband coherent (dashed line) AR and Root-MUSIC algorithms

almost not change with the window gap. Both coherent and non-coherent dualband algorithms shows this effect, where the coherent and the Root-MUSIC approaches have this characteristic more accentuated.

The dualband non-coherent algorithms have a similar performance—with equal total number of samples—as the corresponding single-band approaches. The single-band are slightly better. Actually, the data matrix in the single-band case can be built with more rows—as observed in equations (3.18), (7.5) and (7.17)—and the resulting estimate for the autocorrelation matrix is more accurate. The dualband non-coherent approaches are therefore robust methods to use efficiently all the available radar scenario data observed at different

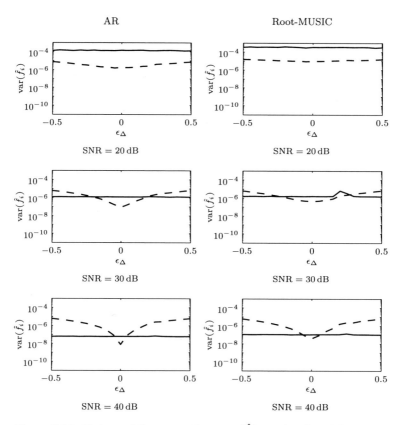

Figure 9.10: Variance of the range estimate $\text{var}(\hat{f}_i)$ as a function of the gap error ϵ_Δ in two-tone scenarios with $\Delta f = \frac{0.3}{N_1+N_2}$, sampled in a dualband system with $\Delta N = 0.25 \cdot (N_1 + N_2)$ for dualband non-coherent (solid line) and dualband coherent (dashed line) AR and Root-MUSIC algorithms

subbands. This is confirmed observing the performance insensitivity to coherence errors. No coherence is necessary for the efficient use of all the subband data.

Dualband coherent algorithms present a better performance than their corresponding non-coherent approaches. The problem in this case is the ambiguity. Only for very closely space targets, great band gaps can be used. For distance between targets of 50 % of the Rayleigh bound, target split problems appear for window gap equal to the size of a subwindow. This effect was also observed when the *discrete Fourier transform* (DFT) is applied to dualband sequences. Sharper peaks are obtained, i. e. a better resolution and accuracy performance can be expected, but further sharp peaks appear at distances which reduce with increasing window gap. A similar coupling between resolution and ambiguity is observed in a study [32] of the ambiguity function of the PANDORA system, a SFCW radar with high synthetic bandwidth and band gaps in-between [23, 24, 33]. A dual effect can also be observed when using a coherent train of pulses (in time domain) as radar signal. With greater time distance between the pulses, both Doppler resolution and maximum unambiguous speed increase [52, 63, 83, 84].

10. Summary and conclusions

The aim of this thesis is to analyze the possibilities of increasing accuracy and resolution performance in multiband radar systems by exploiting the frequency band gap. The general multisensor platform is reduced for the analysis to a simplified dualband radar system, i.e. two radar sensors observing the same radar scenario, with the same aspect angle but operating at different frequency bands. In the analysis, diverse aspects of dualband radar are covered: model for the scattering signal, signal processing, theoretical performance bounds and performance behavior patterns. Special attention is given to the non-coherent case, where the band gap between the subbands is only known approximately.

In order to analyze and develop multiband radar systems, close-to-reality models for the scattering signal are required. A complex scatterer or a group of them can be modeled as a set of individual scattering points with constant delay and frequency dependent amplitude, which in frequency domain are represented by complex oscillations. This model is based on high-frequency approximations for the propagation and scattering of electromagnetic waves, as *geometrical optics* (GO) theory or *geometrical theory of diffraction* (GTD). It has been validated with theoretical, simulated and measured scattering responses of different objects. The delay associated to each scattering center remains constant with frequency, and its amplitude presents a maximum decay with frequency of 25 $^{dB}/_{decade}$. In radar signals sampled in the frequency domain, each scattering center is represented by a pole, whose phase corresponds to the range and its amplitude to the decay with frequency.

As it becomes clear from the signal model, detecting and ranging objects from a radar measurement in the frequency domain is equivalent to detection and frequency estimation of complex oscillations from time-domain collected data. Hence, spectral analysis techniques can be used. Chapter 3 gives a short overview and focuses on the so-called high-resolution *autoregressive* (AR) and MUSIC algorithms, which are the basis for the multiband algorithms developed later.

To assess the possibilities of multiband radar systems, theoretical performance bounds and behavior patterns are of interest. Estimation and detection theories are presented in chapter 4 to address these problems. A powerful tool in estimation theory is the *Cramer-Rao lower bound* (CRLB), which places a lower bound on the estimate variance of any unbiased estimator. Since resolution and accuracy are strongly related to each other—an accurate estimation can only be achieved when the observed targets have been independently detected—the CRLB has been used by many authors to assess resolution bounds. In this thesis, the CRLB estimation theory is only used for accuracy analysis. Resolution is considered a detection problem, i. e. two targets present in a signal are resolved when both are detected, independent of how accurate they are ranged. Detection theory, also called *hypothesis testing* theory, permits the use of an exact and suitable definition for the resolution event and avoids the use of previous knowledge about the number of parameters, i. e. targets, included in the calculation of the CRLB. The obtained results for the single-band case are coherent with those of classical theories, so that its suitability is confirmed. Hypothesis testing theory results in a novel, algorithm-independent and powerful tool to assess resolution bounds.

Theoretical bounds for range accuracy in coherent and non-coherent multiband systems are shown in chapter 5. An increase in range accuracy for increasing band gap between the frequency bands is observed. In coherent systems this improvement is observed for closely and widely spaced targets. In non-coherent systems, where the band gap is an additional parameter to estimate, the accuracy improves only when a high enough number of widely spaced targets—with distances between them over the Rayleigh bound—are present in the signal. The presence of widely spaced targets permits to estimate accurately enough the band gap, so that the range accuracy of the closely spaced targets also increases. A similar performance as in the coherent case is achieved. With increasing band gap, the number of widely spaced targets required to equal the estimate variances of the non-coherent and the coherent systems becomes higher.

In chapter 6, range resolution bounds and behavior patterns for dualband radar systems are obtained with hypothesis testing theory. The probability of resolution P_R as a function of the band gap shows an oscillating behavior for closely spaced targets. The period of this oscillation is inversely proportional to the distance between the two targets present in the signal. This effect is less defined in the non-coherent case, where the oscillations in the P_R are present but with lower intensity. For distances over the Rayleigh bound, the resolution capability is independent from the band gap. The used test functions are the so-called clairvoyant detectors. They assume an unavailable knowledge of the signal's probability density function and give an upper bound in the resolution performance. The resolution in specific radar systems typically falls below the clairvoyant detector resolution limit, although their qualitative behaviors would be similar. The obtained results represent a behavior pattern and lead to the conclusion that no general improvement of the resolution performance in dualband radar systems is expected with increasing band gap.

Coherent and non-coherent multiband AR and MUSIC spectral estimators have been developed in chapter 7. Coherent approaches use the relation between data samples collected at different subbands. As the band gap information is used, the algorithms can benefit from the accuracy improvement predicted in the theory for increasing band gap. However, coherence between the subbands is required. Coherent algorithms suffer also from ambiguity problems, which become more acute with increasing band gap. Non-coherent algorithms do not use the band gap information, and therefore, no coherence is required. Due to the structure of the involved data matrices in the algorithms, the performance of the proposed dualband non-coherent spectral estimators equals that of equivalent single-band estimators with the same total number of available samples. Therefore, non-coherent algorithms result in efficient approaches to exploit all the available data, i. e. bandwidth, in multiband radar systems. A method to combine coherent and non-coherent approaches has been proposed to obtain robustness and enhanced accuracy performance.

The CRLB analysis in chapter 5 has shown the possibilities of non-coherent dualband systems to estimate the unknown band gap, and to obtain the same accuracy performance as coherent systems, when a high enough number of widely spaced targets is present in the radar signal. In chapter 8 one approach to estimate the band gap is presented. The method exploits the relation between the target amplitudes at different subbands to obtain an estimate of the coherence error. Monte Carlo analyses show the unbiased characteristic of the estimator and the growth of the estimate variance with increasing band gap, as predicted by theory.

A Monte Carlo analysis for the resolution and accuracy performance of dualband spectral estimators is presented in chapter 9. The behavior as function of the band gap predicted with the CRLB and the hypothesis testing theory has been confirmed. The resolution performance does not improve continuously with increasing band gap. Instead, a periodic pattern for P_R is obtained for closely spaced targets. The period of this oscillation is inversely proportional to the distance between the two targets present in the signal. For target separation over the Rayleigh bound, P_R can be considered independent from the band gap. Range estimate variance for coherent approaches shows a general decrease with increasing band gap. A periodic pattern as for P_R is also observed. Both coherent and non-coherent dualband algorithms show this effect, where the coherent and the Root-MUSIC approaches have this characteristic more accentuated.

Dualband coherent algorithms present a better performance than their corresponding non-coherent approaches. The problem in this case is the ambiguity. Only for very closely spaced targets, a benefit from great band gaps can be obtained. For a target separation of 50 % of the Rayleigh distance, target split problems appear for window gaps equal to the size of a subwindow.

The dualband non-coherent algorithms have an almost equal performance as equivalent single-band approaches. The dualband non-coherent approaches are therefore robust methods to use efficiently all the available radar scenario data observed at different subbands. This is confirmed by observing the performance insensitivity to coherence errors. No coherence is necessary for the efficient use of all the available subband data.

A. Brief introduction to ray based high frequency approximations

The *Geometrical Optics* (GO) theory and its expansions, the *Geometrical Theory of Diffraction* (GTD) or the *Uniform Geometrical Theory of Diffraction* (UTD), are very intuitive theories which describe the propagation of high frequency electromagnetic waves and their interaction with scattering objects in terms of rays. A comprehensive treatment of these theories can be found in many books [19, 27, 31, 58]. Here only a brief introduction will be given.

A.1. Geometrical Optics

The *Geometrical Optics* (GO) theory is one of the most intuitive approximation techniques to model high-frequency phenomena, i.e. the fields under consideration are in a system where the properties of the medium and scatterer size parameters vary little over an interval on the order of a wavelength. The GO theory starts from the Luneberg-Kline asymptotic expansion for a high-frequency electromagnetic field with a circular frequency ω, in a source free region occupied by an isotropic medium with constitutive parameters ϵ and $\mu = \mu_0$:

$$\vec{E}(\vec{r}, \omega) = e^{-jk\Psi(\vec{r})} \sum_{n=0}^{\infty} \frac{\vec{E}_n(\vec{r})}{(j\omega)^n} \tag{A.1}$$

where $\vec{r}$ is the position vector of the field point, $k^2 = \omega^2 \mu_0 \epsilon$ and $\Psi(\vec{r})$ is the so-called phase function. As the frequency tends to infinity only the first term ($n = 0$) remains, the so called *geometrical optics* field. Substituting (A.1) into the Maxwell's equations and after some manipulation the following equations are obtained:

$$|\vec{\nabla}\Psi| = 1 \tag{A.2}$$

and

$$2(\vec{\nabla}\Psi \cdot \vec{\nabla})\vec{E}_0 + (\vec{\nabla}^2\Psi)\vec{E}_0 = 0 \tag{A.3}$$

the *eikonal* and the *zero*-th order *transport* equations, respectively. The surface of constant Ψ describe wavefronts and the family of all wavefronts describe a system of rays. The rays are everywhere orthogonal to the wavefronts in an isotropic medium and they fulfill the Fermat's principle, which states that the

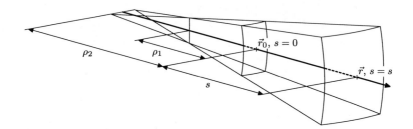

Figure A.1: Infinitesimally narrow diverging astigmatic ray tube

ray path in travel from point a to point b is one for which the optical path length F is stationary.

$$F = \int_a^b \sqrt{\epsilon_r} \mathrm{d}s \tag{A.4}$$

In homogeneous media the rays are straight lines. Integrating the zero-th order transport equation (A.3) from some reference point $\vec{r}_0$ to $\vec{r}$ yields the GO field:

$$\vec{E}(\vec{r}) = \vec{E}_0(\vec{r}_0)e^{-jk\Psi(\vec{r}_0)}\sqrt{\frac{\rho_1\rho_2}{(\rho_1+s)(\rho_2+s)}}e^{-jks} \tag{A.5}$$

where ρ_i are the principal radii of curvature of the wavefront at the reference point $\vec{r}_0$ and s is the optical length from $\vec{r}_0$ to $\vec{r}$. The square root quantity is the ray *divergence factor* which indicates the manner in which the energy spreads along the ray path; it is consequence of the conservation of energy. Next, $\vec{\nabla}\vec{E} = 0$ leads to $\hat{s} \cdot \vec{E}_0 = 0$, which implies that the GO field is polarized transverse to the ray direction $\hat{s}$. The $n = 0$ term in the Luneberg-Kline expansion for the magnetic field intensity is $\vec{H} = \hat{s} \times Y_0\vec{E}$ in which Y_0 is the characteristic admittance of the medium. This relations give the GO ray a local plane wave nature.

A GO ray with both radii of curvature finite and greater than zero is shown in figure A.1, a so called *astigmatic* ray tube, as $\rho_1 \neq \rho_2$. It can be observed that at the distances $s = \rho^i$, as the area of the ray tube becomes zero—in order to conserve the energy transported by the ray—the amplitude of the field should be infinite. This places, where GO predicts an infinite field—which is not physical, and therefore, other methods to predict the field as equivalent currents [40, 70] should be used—are called *caustics*.

For the special case of equal radii of curvature, by setting the reference point at the caustic ($\rho_i \to 0$) the expression for the GO becomes the one of a spherical wavefront:

$$\vec{E}(\vec{r}) = \vec{E}_0(\vec{r}_0)e^{-jk\Psi(\vec{r}_0)}\frac{e^{-jks}}{s}. \tag{A.6}$$

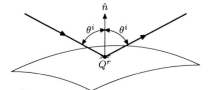

Figure A.2: Reflected ray on a smooth surface

A cylindrical wavefront can also be described in terms of GO rays. If one radii of curvature tends to infinity—as in the case for two dimensional problems—the expression for the ray becomes

$$\vec{E}(\vec{r}) = \vec{E}_0(\vec{r}_0)e^{-jk\Psi(\vec{r}_0)}\sqrt{\frac{\rho}{\rho+s}}e^{-jks} \tag{A.7}$$

and setting the reference point at the caustic line, results in

$$\vec{E}(\vec{r}) = \vec{E}_0(\vec{r}_0)e^{-jk\Psi(\vec{r}_0)}\frac{e^{-jks}}{\sqrt{s}} \tag{A.8}$$

which is the expression of a cylindrical wave. To describe a plane wave, let both radii of curvature go to infinity, so that the ray results in

$$\vec{E}(\vec{r}) = \vec{E}_0(\vec{r}_0)e^{-jk\Psi(\vec{r}_0)}e^{-jks}. \tag{A.9}$$

Reflection

In the GO theory the interaction of electromagnetic waves with arbitrary objects is modeled as reflection or refraction of rays at specular points. Next, only perfectly conducting objects will be considered, where only reflected rays exist. GO analysis of scattering on other media can be found in the literature [19, 31, 42].

One of the principles of the GO theory is the local nature of the high-frequency effects, i.e. the properties of the reflected ray depends only on the surrounding area at the reflection point and on the incident ray.

From the boundary condition $\hat{n} \times [\vec{E}^i + \vec{E}^r] = 0$ on a perfectly conducting smooth surface, where $\hat{n}$ is the surface normal at the point of reflection Q^r (see figure A.2), $\vec{E}^i$ is the incident field and $\vec{E}^r$ the reflected field, it is obtained the following expression for the reflected field:

$$\vec{E}^r(\vec{r}) = \vec{E}^i(Q^r)R_{s,h}\sqrt{\frac{\rho_1^r\rho_2^r}{(\rho_1^r+s)(\rho_2^r+s)}}e^{-jks^r} \tag{A.10}$$

with

$$R_{s,h} = \mp 1. \tag{A.11}$$

The subscripts s and h stand for the acoustic soft $\vec{E}\cdot\hat{n} = 0$ and hard $\partial\vec{E}/\partial n = 0$ conditions, which apply to E-field perpendicular to the plane of incidence or parallel to the plane of incidence respectively. The terms ρ_i^r are the radii of curvature of the reflected field at the reflection point Q^r—which depend only on the radii of curvature of the incident ray and on the surface's geometry at Q^r—and s^r is the path length along the reflected ray. The reflected ray fulfills the Snell's law, i. e. the incident and the reflected angles are equal.

With GO the scattering of conducting objects is modeled with rays reflecting at one or more smooth surfaces and traveling to the observer. GO does not model diffraction effects caused by sharp surfaces like edges, tips, corners,... and also fails to predict the field in shadow regions, i. e. where no ray is present, and therefore, the predicted field is zero. The aim of extensions to the GO, like the *Geometrical Theory of Diffraction* (GTD) or the *Uniform Theory of Diffraction* (UTD) is to overcome this problem by adding diffracted rays, which are also present in the shadow regions.

A.2. Geometrical Theory of Diffraction

The *Geometrical Theory of Diffraction* (GTD) is an extension to the GO with the aim to overcome its limitations by adding additional ray types at sharp surfaces like edges, tips, corners,... and also the so called *creeping rays* or *surface diffracted rays* at smooth surfaces. The diffracted rays, which will also be present in the shadow region predicting there a non-zero field, are expressed as

$$\vec{E}^d(\vec{r}) = \vec{E}^i(Q^d)D\sqrt{\frac{\rho_1^d\rho_2^d}{(\rho_1^d + s)(\rho_2^d + s)}}e^{-jks^d} \tag{A.12}$$

with D being the diffraction coefficient, ρ_i^d the radii of curvature of the diffracted ray and s^d the path length along the diffracted ray. As in GO, the properties of the diffracted ray depends only on the surrounding area at the diffraction point and on the incident ray.

The diffraction coefficients are obtained by analyzing the so called canonical problems, i. e. scattering problems for which an exact solution is available. The expression for the scattered field is rewritten in terms of GO rays to identify the diffraction coefficients. In the following subsections, some canonical scattering problems for perfectly conducting materials will be analyzed. Analysis of other cases and with other materials can be found in the literature [19, 31, 58].

Diffraction at a straight wedge

The general case of a ray diffracted at a straight wedge is shown in figure A.3a. Following the development from Keller [41], a plane wave is incident on the

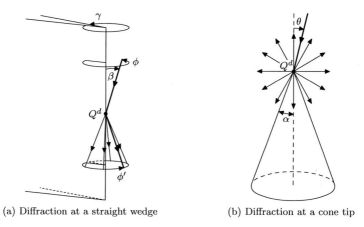

(a) Diffraction at a straight wedge   (b) Diffraction at a cone tip

Figure A.3: Diffracted rays at surface discontinuities

wedge and an infinite set of diffracted rays forming with the wedge the same angle β as the incident ray starts at the diffraction point Q^d.

The field along one of the diffracted rays at the wedge can be expressed as a cylindrical wave with origin at the wedge:

$$\vec{E}^d(\vec{r}) = \vec{E}^i(Q^d)D_{s,h}\frac{e^{-jks^d}}{\sqrt{s^d}} \tag{A.13}$$

where Q^d is the diffraction point, s^d the ray length starting at Q^d and $D_{s,h}$ the diffraction coefficient:

$$D_{s,h} = \frac{e^{-j\pi/4}\sin(\pi/n)}{n\sqrt{2\pi k}\sin\beta}\left[\frac{1}{\cos\left(\frac{\pi}{n}\right) - \cos\left(\frac{\phi-\phi'}{n}\right)} \pm \frac{1}{\cos\left(\frac{\pi}{n}\right) - \cos\left(\frac{\phi+\phi'}{n}\right)}\right] \tag{A.14}$$

with

$$n = \frac{\pi}{\gamma} \tag{A.15}$$

and β, ϕ and ϕ' defined as in the figure A.3a.

Diffraction at a cone tip

The diffraction coefficient for diffraction at cone tips is not well known, especially for directions of incidence far from the axis of the cone. In [18], Felsen

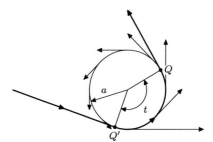

Figure A.4: Creeping rays in a perfectly conducting infinitely long cylinder

gives the solution in terms of GO rays for a plane wave incident on a narrow angle cone with half angle $\alpha \ll \pi/2$, angle of incidence $0 < \theta \leq \alpha$ and soft boundary condition. The diffracted ray can be expressed as a spherical wave with origin at the cone tip,

$$\vec{E}^d(\vec{r}) = \vec{E}^i(Q^d)D_s \frac{e^{-jks^d}}{s^d} \tag{A.16}$$

with Q^d the cone tip, s^d the ray length starting at Q^d, and D_s the diffraction coefficient

$$D_s = \frac{-j}{k} \cdot \frac{3 + \cos^2 \theta}{4 \cos^2 \theta} \left(\frac{\alpha}{2}\right)^2 \qquad \text{for} \quad \alpha \ll \pi/2 \quad \text{and} \quad 0 < \theta \leq \alpha. \tag{A.17}$$

Creeping waves

A creeping ray is *surface-diffracted* ray which propagates along a smooth surface transporting energy to the shadow zone, also in the case where no sharp edges or tips are present. A creeping ray becomes "attached" to a surface at points of grazing incidence, at which it is tangential to the surface, travels along the geodesic path of the object and leaves the object tangentially. This creeping ray spreads its energy in all the directions while propagating along the surface.

For the two dimensional case of a plane wave incident on an infinitely long perfectly conducting cylinder—following the development from Keller [53]—the creeping ray has the expression of a cylindrical wave:

$$\vec{E}^d(\vec{r}) = \vec{E}^i(Q')T_{s,h} \frac{e^{-jks^d}}{\sqrt{s^d}} \tag{A.18}$$

where $\vec{E}^i$ is the GO type incident field at the attachment point Q', s^d is the optical path from the shedding point Q to the observer, and $T_{s,h}$ is the surface diffraction coefficient, which can be expressed as:

$$T_{s,h} = \left(\sum_{n=1}^{N}(D_n^{s,h})^2 e^{-\alpha_n^{s,h}t}\right) e^{-jkt} \tag{A.19}$$

with

$$(D_n^s)^2 = \sqrt{\frac{1}{2\pi k}} \left(\frac{ka}{2}\right)^{\frac{1}{3}} \frac{e^{-j\pi/12}}{[\mathrm{Ai}'(-q_n)]^2}$$
$$(D_n^h)^2 = \sqrt{\frac{1}{2\pi k}} \left(\frac{ka}{2}\right)^{\frac{1}{3}} \frac{e^{-j\pi/12}}{q_n'[\mathrm{Ai}(-q_n')]^2} \tag{A.20}$$

$$\alpha_n^s = \frac{q_n}{a} \left(\frac{ka}{2}\right)^{\frac{1}{3}} e^{j\pi/6}$$
$$\alpha_n^h = \frac{q_n'}{a} \left(\frac{ka}{2}\right)^{\frac{1}{3}} e^{j\pi/6}. \tag{A.21}$$

The quantities $(-q_n)$ are the the zeros of the Airy function Ai, $\mathrm{Ai}(-q_n) = 0$, and $(-q_n')$ the zeros of the derivative of the Airy function Ai', $\mathrm{Ai}'(-q_n') = 0$. Further discussion about the Airy functions can be found in the literature [1, 58].

A.3. Scattering on cylinders using the UTD

In this section, the scattering of a group of infinitely long perfectly conducting cylinders is presented. The used approximation is the *Uniform Theory of Diffraction* (UTD), an improved expansion of the GTD. The UTD is also a ray based approximation with improved reflection and diffraction coefficients—including the Fock Theory [69, 58]—making the solution valid also where GTD fails, i. e. near the shadow boundaries and near the scatterer surfaces. No further general discussion about the UTD is done here, only the coefficients required to the solution of the two-dimensional scattering problem on cylinders are presented. More information about the UTD method and its application to other cases can be found in the literature [27, 58].

A.3.1. UTD coefficients

For the smooth, two-dimensional objects considered here two scattering effects are of concern: reflection and surface diffraction. The reflected ray can be expressed as

$$\vec{E}^r(\vec{r}) = \vec{E}^i(Q^r) R_{s,h} \sqrt{\frac{\rho^r}{\rho^r + s^r}} e^{-jks^r}. \tag{A.22}$$

The term ρ^r is the radii of curvature of the reflected field at the reflection point Q^r and has the expression:

$$\frac{1}{\rho^r} = \frac{1}{\rho^i(Q^r)} + \frac{2}{a_0(Q^r)\cos(\theta^i)} \tag{A.23}$$

where ρ^i is the radii of curvature of the incident field at point Q^r, a_0 is the principal radii of curvature of the surface at point Q^r and θ^i is the angle of incidence. The reflection coefficient is defined as:

$$R_{s,h} = -\sqrt{\frac{-4}{\xi_p}} e^{-j(\xi_p)^3/12} \left\{ \frac{e^{-j\pi/4}}{2\xi_p\sqrt{\pi}}[1 - F(X_p)] + \hat{P}_{s,h}(\xi_p) \right\} \tag{A.24}$$

where $F(x)$ is the *transition function*:

$$F(x) = 2j\sqrt{x}e^{jx} \int\limits_{\sqrt{x}}^{\infty} e^{-ju^2}\,\mathrm{d}u \tag{A.25}$$

$\hat{P}_{s,h}$ is the *Pekeris caret function*:

$$\hat{P}_{s,h}(x) = \frac{e^{-j(\pi/4)}}{\sqrt{\pi}} \int\limits_{-\infty}^{\infty} \frac{\widetilde{Q}V(\tau)}{\widetilde{Q}W_2(\tau)} e^{-jx\tau}\,\mathrm{d}\tau \quad \text{with} \quad \widetilde{Q} = \begin{cases} 1 & \text{soft case} \\ \frac{\partial}{\partial\tau} & \text{hard case} \end{cases} \tag{A.26}$$

where the Fock type Airy functions $V(\tau)$ and $W_2(\tau)$ are defined in [58] where also plots of F and $\hat{P}_{s,h}$ are given. Furthermore the following parameters are defined, the *distance parameter*:

$$L_p = \frac{s^r s^i}{s^r + s^i} \tag{A.27}$$

the parameter of the *transition function*:

$$X_p = 2kL_p \cos^2(\theta^i) \tag{A.28}$$

the *Fock parameter*:

$$\xi_p = -2m(Q^r)\cos\theta^i \tag{A.29}$$

and $m(Q^r)$, the *curvature* parameter:

$$m(Q^r) = \left[\frac{ka_0(Q^r)}{2} \right] \tag{A.30}$$

It can be shown [58] that the UTD expression of $R_{s,h}$ for a point in the deep lit region (field point far away from the shadow boundaries) reduces to the one obtained in the GO, $R_{s,h} = \mp 1$.

The surface diffraction ray has the expression:

$$\vec{E}^d(\vec{r}) = \vec{E}^i(Q')T_{s,h}e^{-jks^d}\frac{e^{-jks^d}}{\sqrt{s^d}} \tag{A.31}$$

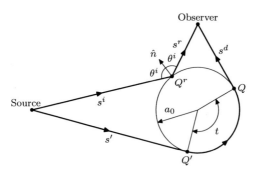

Figure A.5: Reflected and diffracted rays at a cylinder

where $\vec{E}^i$ is the GO type incident field at attachment point Q', s^d the distance from shedding point Q to the observer (see figure A.5) and $T_{s,h}$ has the expression:

$$T_{s,h} = -\sqrt{m(Q')m(Q)}\sqrt{\frac{2}{k}}e^{-jkt}\left\{\frac{e^{-j\pi/4}}{2\xi_d\sqrt{\pi}}\left[1 - F(X_d)\right] + \hat{P}_{s,h}(\xi_d)\right\}. \quad (A.32)$$

$T_{s,h}$ includes the integrated effect of the surface curvature over the entire path $Q'Q$ through its argument ξ_d:

$$\xi_d = \int_{\tau(Q')}^{\tau(Q)} \frac{m(\tau)}{a_0(\tau)}\mathrm{d}\tau \quad (A.33)$$

with

$$m(\tau) = \left[\frac{ka_0(\tau)}{2}\right]^{1/3}. \quad (A.34)$$

As earlier, $a_0(\tau)$ is the radius of curvature at some general point located by the parameter τ. The quantity

$$t = \int_{\tau(Q')}^{\tau(Q)} \mathrm{d}\tau \quad (A.35)$$

is the path (or arc) length traversed by the surface ray between Q' and Q, and effects $T_{s,h}$ directly through the phase factor e^{-jkt}, but it is not the only phase shift added by $T_{s,h}$. As earlier the *distance* parameter is defined as:

$$L_d = \frac{s^d s'}{s^d + s'} \quad (A.36)$$

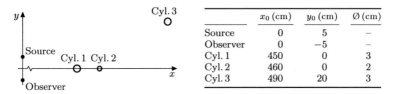

	x_0 (cm)	y_0 (cm)	Ø (cm)
Source	0	5	–
Observer	0	−5	–
Cyl. 1	450	0	3
Cyl. 2	460	0	2
Cyl. 3	490	20	3

Figure A.6: Cylinder scenarios simulated with the UTD. The first scenario includes cylinder 1, the second cylinders 1 and 2, and the third cylinders 1, 2 and 3.

and

$$X_d = \frac{kL_d(\xi)^2}{2m(Q')m(Q)}. \tag{A.37}$$

For a point located in the deep shadow region, far away from the shadow boundary, it can be shown [58] that the UTD solution coincides with the one predicted by the GTD.

Due to the surface diffracted rays the field solutions obtained with the UTD are much more accurate than those obtained with GO. Also, objects at lower frequencies, down to $k_0 a = 2$ where a is the minimum radius of the object along the creeping ray trajectories, can be modeled [69]. The interaction of different objects with each other is also modeled accurately with the UTD approximation, thanks again to the creeping rays and to the accuracy in the solutions of rays near the surfaces, which are the incidents ones at the surrounding objects, which back-reflect or back-diffract the rays to the observer.

A.3.2. Simulation results

Three two-dimensional cylinder scenarios have been simulated with UTD and also with GO. The scenarios consist of a source line, an observation point and a set of perfectly conducting cylinders as shown in figure A.6, where the cylinders and the source line are infinitely long in the z-direction. The radiated E-field is perpendicular to the cylinder axis, i. e. the "hard" boundary condition has been considered.

One of the most time-consuming process when using the UTD to calculate the scattering of a given scenario is the ray-tracing process: obtaining all the possible rays traveling from the source to the observer fulfilling the Fermat's principle (A.4). For the cylinder scenarios described above a ray-tracing algorithm has been developed.

The algorithm searches for rays starting at the source line or tangentially at the cylinder surfaces and reaching, allowing a limited number of reflections, the observer or a tangent point at one of the cylinders. Properly combining this rays (a diffracted ray is the combination of one ending at the surface and starting again at it) all the possible ray paths are obtained.

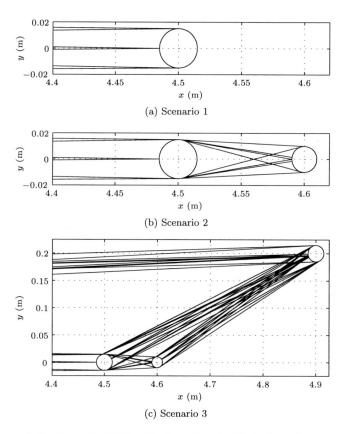

(a) Scenario 1

(b) Scenario 2

(c) Scenario 3

Figure A.7: Ray paths fulfilling the Fermat's principle in the cylinder scenarios described in figure A.6

This rays are obtained by means of a zero-search. For each possible start a set of rays is launched and it is observed which consecutive rays differ in the object they hit at or in the relative position to the observer. Two weighting functions with the start direction as independent variable are defined: one with a zero-cross where the ray is tangent to a cylinder, and other with a zero-cross when it travels through the observer. A zero-search is started within the intervals and with the weighting functions defined previously, and the rays are obtained.

Figure A.7 shows the rays obtained for the cylinder scenarios using this algorithm and allowing a maximum of three reflection and/or diffraction effects.

Figure A.8 shows the scattered field in time-domain for the three cylinder scenarios at different frequency bands, obtained with GO, UTD and from measurements with cylinder scenarios of the same xy-plane distribution and 1.5 m longitude as described in appendix B. The measured data has been made coincident to the simulated one at the main peak. The main reflection at the cylinders and the "creeping wave" terms can be clearly identified in the time-domain plots in all the bands. In figure A.8(c) around the 5.3 m position can also be observed the effect of multiple reflections at the cylinders before traveling back to the observer.

Comparing GO with UTD, it is observed that GO fails to model the presence of cylinder no. 2. This cylinder lies in the shadow region of the first one and therefore it is invisible for the GO, which does not include the diffracted rays. It can be also remarked that the interaction of the first and third cylinder, region around 5.3 m, are modeled by GO and UTD very similarly, therefore, it can be concluded that the main contribution to this effect are just the reflected rays and not the diffracted ones.

Comparing the measured with the simulated data, it is seen that in the prediction of the peak positions by UTD there is no error and the amplitude error is lower than 3 dB. In the third scenario at V-band, between the first and second cylinder the measured values are approximately 20 dB over the simulated data. This can be due to the bench where the cylinder where mounted on and to the wood plate located at the top of the cylinder to ensure stability. The same effect is observed between the second and the third cylinder, and before the first one. At 5.3 m the the mutual interaction between the cylinder is seen, also modeled by the UTD.

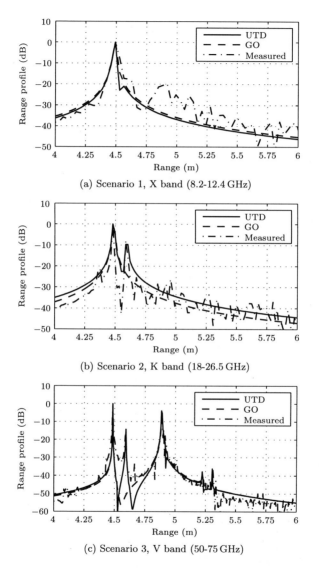

(a) Scenario 1, X band (8.2-12.4 GHz)

(b) Scenario 2, K band (18-26.5 GHz)

(c) Scenario 3, V band (50-75 GHz)

Figure A.8: Range profiles from the cylinder scenarios shown in figure A.6 simulated with UTD, GO and measured

B. Radar measurements description

This appendix describes the equipment, data structure and calibration techniques used in the radar measurements appearing in this thesis. The measurements are realized in an anechoic chamber using laboratory equipment. A radar scenario can be seen as a one-port microwave network where the reflection coefficient is the radar response. Therefore, a radar system can be implemented in the laboratory using a network analyzer with a reflection measurement configuration. The laboratory gives the added advantages of controlled environment and more system design flexibility. The measurements are realized in the frequency domain at discrete frequency points. To obtain a complete set of ultra-wideband measurements three widely spaced frequency bands have been selected for the measurements: X (8.2-12.4 GHz), K (18-26.5 GHz) and V (50-75 GHz).

B.1. Description of the measurement equipment

Block diagrams of the measurement equipment are shown in figures B.1 and B.2. Depending on the frequency band, different hardware configurations are required.

Baseband configuration

The block diagram in figure B.1 shows the baseband configuration, which is used for the measurements in X and K bands.

The synthesizer generates the RF test signal, which after amplification—only for X band—by a power amplifier is radiated by the source antenna. This radiated signal arrives the receiving antenna after backscattering at the test target.

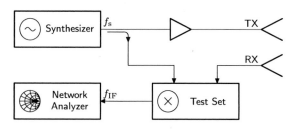

Figure B.1: Block diagram of the measurement equipment in the baseband configuration

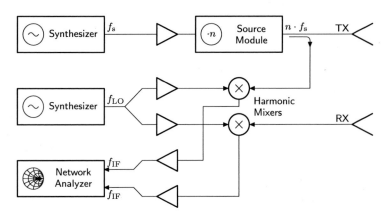

Figure B.2: Block diagram of the measurement equipment in the mm-wave-band configuration

A directional coupler connected immediately after the synthesizer separates a portion of the transmitted signal to the port a_1 of the test set, where the receiving antenna is also connected to the port b_1. A conversion of the RF signals to an intermediate frequency (IF) of 20 MHz takes place in it.

The resulting IF signals are applied to the network analyzer for detection, post-processing, and display. A computer connected to the analyzer saves the reflection coefficient $s_{11} = b_1/a_1$.

mm-Wave-band configuration

A block diagram of the mm-wave-band configuration is shown in figure B.2. It is used for the measurements in V Band.

The stimulus for the source antenna is generated using the synthesized source followed by a power amplifier and a frequency multiplier. The directional coupler separates a portion of the transmitted signal to the harmonic mixer for the IF conversion. Connected to the receiving antenna is also an harmonic mixer for the conversion of the received RF signal to the IF signal.

The second source provides the LO for all of the harmonic mixers. This LO source is set such that the mm-wave test signal frequency and the appropriate LO harmonic are offset by exactly 20 MHz.

The IF signals are amplified and then applied to the network analyzer for detection, post-processing, and display. A computer connected to the analyzer saves the reflection coefficient s_{11}.

Table B.1: Parameters of the measured datasets

Band	N	Start (GHz)	Δf (MHz)	End (GHz)	R_{unamb} (m)	Δr (mm)
X	1 201	8.2	3.5	12.4	42.8275	35.67
K	2 429	18	3.5	26.498	42.8275	17.64
V	7 143	50	3.5	74.997	42.8275	5.77

B.2. Dataset characteristics

The RF equipment measures the complex reflection coefficient in the frequency domain at N discrete frequency points separated by Δf, i. e. using a *stepped frequency continuous wave* (SFCW) radar approach. The frequency step Δf fixes the radar unambiguous range

$$R_{\text{unamb}} = \frac{c_0}{2\Delta f} \tag{B.1}$$

where c_0 is the light speed, and the total number of frequency points, i. e. the total bandwidth, the achievable resolution

$$\Delta r = \frac{c_0}{2\Delta f N}. \tag{B.2}$$

In table B.1 the most important dataset parameters are summarized.

B.3. Calibration method

For the development of the calibration method the error model depicted in figure B.3 is assumed. The reflection coefficient measured with the system r_M is the one of the target under test r_{target} buried in the background and viewed through the interference quadripole **S**. The interference quadripole models the measuring equipment: the cables, the amplifiers, the antennas, the coupling between transmission and reception,... The background models all the effects associated with the non ideal free space environment, reflections at the walls of the chamber, at the floor, at residual objects,...

The expression of the measured reflection coefficient r_M assuming no interaction of the background with the target under test can be expressed as:

$$r_M = s_{11} + \frac{s_{12}s_{21} \cdot (r_{\text{target}} + \text{Background})}{1 - s_{22} \cdot (r_{\text{target}} + \text{Background})}. \tag{B.3}$$

It is reasonable to assume that $|s_{22} \cdot (r_{\text{target}} + \text{Background})| \ll 1$ (s_{22} models the back-reflection of the measuring system to the target) is fulfilled, and therefore, the expression can be reduced to:

$$r_M = s_{11} + s_{12}s_{21} \cdot (r_{\text{target}} + \text{Background}) \tag{B.4}$$

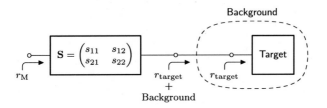

Figure B.3: Error model assumed for the calibration technique

The proposed calibration technique uses the measurements of the chamber without object r_{chamber} and with a target with constant and well defined reflection coefficient, e.g. a corner reflector $r_{\text{reflector}}$. The calibration is based on the suppression of the s_{11} and the background effects by the subtraction of the empty chamber measurement, and of the interference quadripole $s_{12}s_{21}$ effect by the normalization with the corner reflector:

$$r_{\text{C}} = \frac{r_{\text{M}} - r_{\text{chamber}}}{r_{\text{reflector}} - r_{\text{chamber}}} \tag{B.5}$$

Identifying the r_{chamber} and $r_{\text{reflector}}$ terms as:

$$r_{\text{chamber}} = r_{\text{M}}|_{r_{\text{target}}=0} = s_{11} + s_{12}s_{21} \cdot \text{Background} \tag{B.6}$$

$$r_{\text{reflector}} = r_{\text{M}}|_{r_{\text{target}}=1} = s_{11} + s_{12}s_{21} \cdot (1 + \text{Background}) \tag{B.7}$$

it is easy to see that the result is $r_{\text{C}} = r_{\text{target}}$.

C. Glossary of abbreviations

ACS	autocorrelation sequence
AIC	Akaike information criterion
AR	autoregressive
ARMA	autoregressive-moving average
CAT	criterion autoregressive transfer
CRLB	Cramer-Rao lower bound
CW	continuous wave
CWGN	complex white Gaussian noise
DFT	discrete Fourier transform
ESPRIT	estimation of signal parameters via rotational invariance techniques
FIR	finite impulse response
FMCW	frequency modulated continuous wave
FSK-CW	frequency shift keying continuous wave
GO	geometrical optics
GTD	geometrical theory of diffraction
LRT	likelihood ratio test
MA	moving average
MAP	maximum a posteriori
MCPC	multicarrier phase-coded
MDL	minimum description length
ML	maximum likelihood
MUSIC	multiple signal classification
OFDM	orthogonal frequency division multiplexing
PDF	probability density function
PHD	Pisarenko harmonic decomposition
PSD	power spectral density
PSF	point-spread function
SFCW	stepped frequency continuous wave
SNR	signal-to-noise ratio
UTD	uniform theory of diffraction

Bibliography

[1] Abramowitz, M. and I. A. Stegun: *Handbook of Mathematical Functions*. Dover Publications, Inc., New York, 1972.

[2] Akaike, H.: *Power Spectrum Estimation through Autoregression Model Fitting*. Annals of the Institute of Statistical Mathematics, 21:407–419, 1969.

[3] Akaike, H.: *A New Look at the Statistical Model Identification*. IEEE Transactions on Automatic Control, 19(6):69–72, December 1974.

[4] Balanis, C. A.: *Advanced Engineering Electromagnetics*. John Wiley & Sons, New York, 1989.

[5] Barabell, A. J.: *Improving the Resolution Performance of Eigenstructure-Based Direction-Finding Algorithms*. In *IEEE International Conference on Acoustics, Speech, and Signal Processing*, volume 8, pages 336–339, Boston, USA, April 1983.

[6] Bartlett, M. S.: *Smoothing Periodograms from Time Series with Continuous Spectra*. Nature, 161:686–687, May 1948.

[7] Blackman, R. B. and J. W. Tukey: *The Measurement of Power Spectra from the Point of View of Communication Engineering*. Dover Publications, Inc., New York, 1958.

[8] Borovikov, V. A.: *Geometrical Theory of Diffraction*. The Institution of Electrical Engineers, Stevenage, Herts, UK, 1994.

[9] Bronstein, I. N., K. A. Semendjajew, G. Musiol und H. Mühlig: *Taschenbuch der Mathematik*. Verlag Harri Deutsch, Thun und Frankfurt am Main, 5. überarbeitete und erweiterte Auflage, 2001.

[10] Burg, J. P.: *Maximum Entropy Spectral Analysis*. In *Proceedings of the 37th Meeting of the Society of Exploration Geophysicists*, Oklahoma City, USA, October 1967.

[11] Burg, J. P.: *Maximum Entropy Spectral Analysis*. PhD thesis, Department of Geophysics, Standford University, Standford, USA, May 1975.

[12] Cook, C. E. and M. Bernfeld: *Radar Signals. An Introduction to Theory and Applications.* Academic Press, New York, London, 1967.

[13] Cornbleet, S.: *Microwave and Geometrical Optics.* Academic Press, London, 1994.

[14] Cuomo, K. M., J. E. Piou, and J. T. Mayhan: *Ultrawide-Band Coherent Processing.* IEEE Transactions on Antennas and Propagation, 47(6):1094–1107, June 1999.

[15] Daniell, P. J.: *On the Theoretical Specification and Sampling Properties of Autocorrelated Time-Series.* Journal of the Royal Statistical Society, Series B, 8:88–90, 1946.

[16] Detlefsen, J.: *Radartechnik: Grundlagen, Bauelemente, Verfahren, Anwendungen.* Springer, Berlin, 1989.

[17] Farrier, D. R., D. J. Jeffries, and R. Mardani: *Theoretical Performance Prediction of the MUSIC Algorithm.* IEE Proceedings on Radar and Signal Processing, 135(3):216–222, June 1988.

[18] Felsen, L. B.: *Plane-Wave Scattering by Small-Angle Cones.* IRE Transactions, 5:211, 1957.

[19] Felsen, L. B. and N. Marcuvitz: *Radiation and Scattering of Waves.* Prentice Hall, Englewood Cliffs, New Jersey, 1973.

[20] Friedlander, B.: *A Sensitivity Analysis of the MUSIC Algorithm.* IEEE Transactions on Acoustics, Speech and Signal Processing, 38(10):1740–1751, October 1990.

[21] Fuchs, J. J.: *Estimating the Number of Sinusoids in Additive White Noise.* IEEE Transactions on Acoustics, Speech and Signal Processing, 36(10):1846–1853, December 1988.

[22] Gabor, D.: *Theory of Communication.* IEE Journal, 93(Nr. pt. III):429–441, 1946.

[23] Genderen, P. van: *Multi-Waveform SFCW radar.* In *33rd European Microwave Conference*, pages 849–852, Munich, Germany, October 2003.

[24] Genderen, P. van and I. Nicolaescu: *System Description of a Stepped Frequency CW Radar for Humanitarian Demining.* In *2nd International Workshop on Advanced GPR*, pages 9–15, Delft, Netherlands, May 2003.

[25] Gu, H.: *Frequency Resolution and Estimation of AR Spectral Analysis.* IEEE Transactions on Signal Processing, 41(1):432–436, January 1993.

[26] Hall, D. L. and J. Llinas: *An introduction to multisensor data fusion.* Proceedings of the IEEE, 85(1):6–23, January 1997.

[27] Hansen, R. C.: *Geometric Theory of Diffraction.* IEEE Press, New York, 1981.

[28] Haykin, S. (editor): *Nonlinear Methods of Spectral Analysis.* Springer, Berlin, Heidelberg, New York, 1979.

[29] Haykin, S., W. Harrison, and L. Bamber: *Adaptive Filter Theory.* Prentice Hall, Upper Saddle River, New Jersey, 2003.

[30] Ishimaru, A.: *Electromagnetic Wave Propagation, Radiation, and Scattering.* Prentice Hall, Englewood Cliffs, New Jersey, 1991.

[31] James, G. L.: *Geometrical Theory of Diffraction for Electromagnetic Waves.* Peter Peregrinus, Stevenage, Herts, UK, Revised Third edition, 1986.

[32] Jankiraman, M., E. W. de Jong, and P. van Genderen: *Ambiguity Analysis of PANDORA Multifrequency FMCW/SFCW Radar.* In *The Record of the IEEE 2000 International Radar Conference*, pages 35–41, Alexandria, USA, May 2000.

[33] Jankiraman, M., B. J. Wessels, and P. van Genderen: *PANDORA Multifrequency FMCW/SFCW Radar.* In *The Record of the IEEE 2000 International Radar Conference*, pages 750–757, Alexandria, USA, May 2000.

[34] Johnson, D. H.: *The Application of Spectral Estimation Methods to Bearing Estimation Problems.* Proceedings of the IEEE, 70(9):1018–1028, September 1982.

[35] Kaveh, M. and A. Barabell: *The Statistical Performance of the MUSIC and the Minimum-Norm Algorithms in Resolving Plane Waves in Noise.* IEEE Transactions on Acoustics, Speech, and Signal Processing, 34(2):331–341, April 1986.

[36] Kay, S. M.: *Modern Spectral Estimation.* Prentice Hall, Englewood Cliffs, New Jersey, 1988.

[37] Kay, S. M.: *Fundamentals of Statistical Signal Processing. Vol. I Estimation Theory.* Prentice Hall, Upper Saddle River, New Jersey, 1993.

[38] Kay, S. M.: *Fundamentals of Statistical Signal Processing. Vol. II Detection Theory.* Prentice Hall, Upper Saddle River, New Jersey, 1998.

[39] Kay, S. M. and S. L. Marple: *Spectrum Analysis. A Modern Perspective.* Proceedings of the IEEE, 69(11):1380–1419, November 1981.

[40] Keller, J. B.: *Diffraction by an Aperture.* Journal Applied Physics, 28:426, 1957.

[41] Keller, J. B.: *Geometric Theory of Diffraction.* Journal Optical Society of America, 52:116–130, 1962.

[42] Kouyoumjian, R., L. Jr. Peters, and D. Thomas: *A Modified Geometrical Optics Method for Scattering by Dielectric Bodies.* IEEE Transactions on Antennas and Propagation, 11(6):690–703, November 1963.

[43] Lee, H. B.: *The Cramér-Rao Bound on Frequency Estimates of Signals Closely Spaced in Frequency.* IEEE Transactions on Signal Processing, 40(6):1508–1517, June 1992.

[44] Lee, H. B. and M. S. Wengrovitz: *Resolution Threshold of Beamspace MUSIC For Two Closely Spaced Emitters.* IEEE Transactions on Acoustics, Speech, and Signal Processing, 38(9):1545–1559, September 1990.

[45] Lee, H. B. and M. S. Wengrovitz: *Statistical Characterization of the MUSIC Null Spectrum.* IEEE Transactions on Signal Processing, 39(6):1333–1347, June 1991.

[46] Lee, H. B. and M. S. Wengrovitz: *Theoretical Resolution-Threshold Curve for the MUSIC Algorithm.* In *International Conference on Acoustics, Speech and Signal Processing*, volume 5, pages 3313–3316, Toronto, Canada, April 1991.

[47] Lehmann, E. L.: *Testing Statistical Hypotheses.* John Wiley & Sons, New York, 1959.

[48] Lehmann, E. L. and J. P. Romano: *Testing Statistical Hypotheses.* Springer, Berlin, Heidelberg, New York, Third edition, 2005.

[49] Levanon, N.: *Radar Principles.* John Wiley & Sons, New York, 1988.

[50] Levanon, N.: *Multifrequency Complementary Phase-Coded Radar Signal.* IEE Proceedings on Radar Sonar and Navigation, 147(6):276–284, December 2000.

[51] Levanon, N. and E. Mozeson: *Multicarrier Radar Signal - Pulse Train and CW.* IEEE Transactions on Aerospace and Electronic Systems, 38(2):707–720, April 2002.

[52] Levanon, N. and E. Mozeson: *Radar Signals*. John Wiley & Sons, New York, 2004.

[53] Levy, B. R. and J. B. Keller: *Diffraction by a Smooth Object*. Communications on Pure and Applied Mathematics, 12:159–209, 1959.

[54] Li, S. and B.W. Dickinson: *Performance Contours of Autoregressive Estimates*. IEEE Transactions on Acoustics, Speech, and Signal Processing, 36(4):608–610, April 1988.

[55] Ludloff, A.: *Handbuch Radar und Radarsignalverarbeitung*. Vieweg, Braunschweig, 1993.

[56] Marple, S. L.: *Frequency Resolution of Fourier and Maximum Entropy Spectral Estimates*. Geophysics, 47(9):1303–1307, September 1982.

[57] Marple, S. L.: *Digital Spectral Analysis*. Prentice Hall, Englewood Cliffs, New Jersey, 1987.

[58] McNamara, D. A., C. W. I. Pistorius, and J. A. G. Malherbe: *Introduction to the Uniform Geometrical Theory of Diffraction*. Artech House, Boston, London, 1990.

[59] Meinke, H. H. und H. Groll: *Radar. Physikalische Voraussetzungen und Technische Anwendung*. Reclam, Stuttgart, 1962.

[60] Nuttal, A. H.: *Spectral Analysis of an univariate Process with Bad Data Points, via Maximum entropy and Linear Predictive Techniques*. Technical Report TR-5303, Naval Underwater System Center, New London, USA, March 1976.

[61] Ondiviela, J.: *Validation of an exponential model for the scattering of complex objects*. Diplomarbeit, Lehrstuhl für Hochfrequenztechnik, Technische Universität München, Munich, Germany, May 2005.

[62] Parzen, E.: *Some Recent Advances in Time Series Modeling*. IEEE Transactions on Automatic Control, 19(6):723–730, December 1974.

[63] Peebles, P. Z.: *Radar Principles*. John Wiley & Sons, New York, 1998.

[64] Pisarenko, V. F.: *The Retrieval of Harmonics from a Covariance Function*. Geophysics J.R. Astron. Soc., 33:347–366, 1973.

[65] Proakis, John G., Charles M. Rader, Fuyun Ling, and Chrysostomos L. Nikias: *Advanced Digital Signal Processing*. Macmillan, New York, 1992.

[66] Rao, B. D. and K. V. S. Hari: *Performance Analysis of Root-Music*. IEEE Transactions on Acoustics, Speech and Signal Processing, 37(12):1939–1949, December 1989.

[67] Rissanen, J.: *A Universal Prior for the Integers and Estimation by Minimum Description Length*. Annals of Statistics, 11:417–431, 1983.

[68] Roy, R. and T. Kaylath: *ESPRIT Estimation of Signal Parameters via Rotational Invariance Techniques*. IEEE Transactions on Acoustics, Speech and Signal Processing, 37(7):984–995, July 1989.

[69] Ruck, G. T., D. E. Barrick, W. D. Stuart, and C. K. Krichbaum: *Radar Cross Section Handbook*. Plenum Press, New York, London, 1970.

[70] Ryan, C. E. and L. P. Peters: *Evaluation of Edge-Diffracted Fields Including Equivalent Currents for the Caustic Regions*. IEEE Transactions on Antennas and Propagation, 17(3):292–299, May 1969.

[71] Schmidt, R. O.: *A Signal Subspace Approach to Multiple Emitter Location and Spectral Estimation*. PhD thesis, Department of Electrical Engineering, Stanford University, Standford, USA, November 1981.

[72] Schmidt, R. O.: *Multiple Emitter Location and Signal Parameter Estimation*. IEEE Transactions on Antennas and Propagation, 34(3):276–280, March 1986.

[73] Shahram, M. and P. Milanfar: *Imaging Below the Diffraction Limit: a Statistical Analysis*. IEEE Transactions on Image Processing, 13(5):677–689, May 2004.

[74] Shahram, M. and P. Milanfar: *On the Resolvability of Sinusoids with Nearby Frequencies in the Presence of Noise*. IEEE Transactions on Signal Processing, 53(7):2579–2588, July 2005.

[75] Siart, U.: *Modellgestützte Signalverarbeitung für inkohärente Radarsensoren in mehreren Frequenzbändern*. Doktorarbeit, Fakultät für Elektrotechnik und Informationstechnik, Technische Universität München, Munich, Germany, 2005. Published by Logos Verlag, Berlin.

[76] Siart, U. and S. Tejero: *UTD Modelling of Radar Scenes for Coherent Multiband Processing*. In Russer, P. and M. Mongiardo (editors): *Fields, Networks, Computational Methods, and Systems in Modern Electrodynamics*, Springer Proceedings in Physics. Springer, Berlin, Germany, 2004.

[77] Siart, U., S. Tejero, and J. Detlefsen: *Concepts for Coherent Processing of Multiple Radar Sensor Signals*. In *Workshop WS11 European Microwave Week*, Munich, Germany, October 2003.

[78] Siart, U., S. Tejero, and J. Detlefsen: *Improving Range Resolution by Coherent Subband Processing of Multiple Radar Sensor Data*. In *International Radar Symposium*, pages 101–104, Dresden, Germany, September-October 2003.

[79] Siart, U., S. Tejero und J. Detlefsen: *Kohärente Zweibandsensorik für Radaranwendungen*. In: *Workshop der URSI-Kommission B*, Günzburg, Germany, March 2004.

[80] Siart, U., S. Tejero, and J. Detlefsen: *Resolution Properties of Spectral Estimators Applied to Multiple Frequency Bands*. In *International Symposium on Signals, Systems, and Electronics*, Linz, Austria, August 2004.

[81] Siart, U., S. Tejero, and J. Detlefsen: *Cramér-Rao-Bound for Coherent Dual-Band Radar Range Estimation*. In *German Microwave Conference*, pages 258–261, Ulm, Germany, April 2005.

[82] Siart, U., S. Tejero, and J. Detlefsen: *Exponential Modelling for Mutual-Cohering of Subband Radar Data*. Advances in Radio Science, 3:199–204, 2005.

[83] Skolnik, M. I.: *Introduction to Radar Systems*. McGraw-Hill, New York, 1962.

[84] Skolnik, M. I.: *Introduction to Radar Systems*. McGraw-Hill, New York, Second edition, 1980.

[85] Smith, S. T.: *Statistical Resolution Limits and the Complexified Cramér-Rao Bound*. IEEE Transactions on Signal Processing, 53(5):1597–1609, May 2005.

[86] Srinivas, K. R. and V. U. Reddy: *Finite Data Performance of MUSIC and Minimum Norm Methods*. IEEE Transactions on Aerospace and Electronic Systems, 30(1):161–174, January 1994.

[87] Stoica, P. and Nehorai A.: *MUSIC, Maximum Likelihood, and Cramer-Rao Bound*. IEEE Transactions on Acoustics, Speech, and Signal Processing, 37(5):720–741, May 1989.

[88] Stoica, P. and A. Nehorai: *MUSIC, Maximum Likelihood, and Cramer-Rao Bound: Further Results and Comparisons*. IEEE Transactions on Acoustics, Speech, and Signal Processing, 38(12):2140–2150, December 1990.

[89] Swingler, D. N.: *Narrowband Line-Array Beamforming: Practically Achievable Resolution Limit of Unbiased Estimators*. IEEE Journal of Oceanic Engineering, 19(2):225–226, April 1994.

[90] Tejero, S., U. Siart, and J. Detlefsen: *Modeling and Processing of Multi-band Radar Sensor Data*. In *German Radar Symposium*, pages 491–495, Bonn, Germany, September 2002.

[91] Tejero, S., U. Siart, and J. Detlefsen: *Coherent and Non-Coherent Processing of Multiband Radar Sensor Data*. Advances in Radio Science, 4:73–78, 2006.

[92] Ulrych, T. J. and R. W. Clayton: *Time Series Modeling and Maximum Entropy*. Physics of the Earth and Planetary Interiors, 12:188–200, August 1976.

[93] Wax, M. and T. Kailath: *Detection of signals by Information Theoretic Criteria*. IEEE Transactions on Acoustics, Speech and Signal Processing, 33(2):387–392, April 1985.

[94] Welch, P. D.: *The Use of Fast Fourier Transform for the Estimation of Power Spectra: A Method Based on Time Averaging over Short Modified Periodograms*. IEEE Transactions on Audio and Electroacoustics, 15:70–73, June 1967.

[95] Widrow, B. and S. D. Stearns: *Adaptive Signal Processing*. Prentice Hall, Englewood Cliffs, New Jersey, 1985.

[96] Woodward, P. M.: *Probability and Information Theory, with Applications to Radar*. Pergamon Press, Oxford, 1953.

[97] Ying, C.-H. J., A. Sabharwal, and R. L. Moses: *A Combined Order Selection and Parameter Estimation Algorithm for Undamped Exponentials*. IEEE Transactions on Signal Processing, 48(3):693–701, March 2000.

[98] Zatman, M. and S. T. Smith: *Resolution and Ambiguity Bounds for Interferometric-Like Systems*. In *Conference Record of the Thirty-Second Asilomar Conference on Signals, Systems & Computers*, volume 2, pages 1466–1470, Pacific Grove, USA, November 1998.

[99] Zhang, Q. T.: *Probability of Resolution of the MUSIC Algorithm*. IEEE Transactions on Signal Processing, 43(4):978–987, April 1995.

[100] Zhang, Q. T.: *A Statistical Resolution Theory of the AR Method of Spectral Analysis*. IEEE Transactions on Signal Processing, 46(10):2757–2766, October 1998.

[101] Zhou, C., F. Haber, and D. L. Jaggard: *A Resolution Measure for the MUSIC Algorithm and its Application to Plane Wave Arrivals Contaminated by Coherent Interference*. IEEE Transactions on Signal Processing, 39(2):454–463, February 1991.